# The Logical Design
# of Multiple-Microprocessor
# Systems

**B. A. Bowen and R. J. A. Buhr**

*Department of Systems Engineering and Computing Science*
*Carleton University*

**Prentice-Hall, Inc.**
**Englewood Cliffs, New Jersey 07632**

**Library of Congress Cataloging in Publication Data**

Bowen, B   A
    The logical design of multiple-microprocessor
systems.

    Bibliography:  p.
    Includes index.
    1.  Microprocessors.  2.  System design.  3.  Paral-
lel processing (Electronic computers)  I.  Buhr, R. J.A.,
joint author.  II.  Title.
QA76.5.B664      001.64      80-10787
ISBN 0-13-539908-4

Printed in the United States of America

10 9 8 7 6 5 4 3 2 1

Prentice-Hall International, Inc., *London*
Prentice-Hall of Australia Pty. Limited, *Sydney*
Prentice-Hall of Canada, Ltd., *Toronto*
Prentice-Hall of India Private Limited, *New Delhi*
Prentice-Hall of Japan, Inc., *Tokyo*
Prentice-Hall of Southeast Asia Pte. Ltd., *Singapore*
Whitehall Books Limited, *Wellington, New Zealand*

# Table of Contents

**Chapter 1**

**Classical Concepts and Concurrent Systems**

# Chapter 2


## Processes — The Active Components

# Chapter 3


## Monitors

## Chapter 4

### Hardware — The Inner Circle

# Chapter 5

## The Kernel

# Chapter 6

## How Are Concurrent Systems Designed?

# Chapter 7

## A Multiprocessor Packet Switch

## A: Requirements Definition

# Chapter 8

## Concurrent High Level Languages

# Chapter 9

## Systems Issues

# Preface

This book evolved from experiences gained in academic research and in a laboratory during the design and construction of a distributed intelligent terminal system using multiple microprocessors.

The authors were drawn together from an original interest in hardware and concurrent operating systems. This combination of expertise led to the formulation of the Microprocessor Systems Development Laboratory and the creation of a prototype system. It was the raw experience of creating this system which provided the crucible for testing concepts, principles and design algorithms.

The material fell naturally into three parts. In Part A a tutorial presentation of the key concepts required to describe and understand concurrent systems. This material derives from the wealth of concepts which have been used to design real-time as well as concurrent operating systems on uniprocessors. The beginnings of a design philosophy are threaded through this part and certainly an attitude toward this process is clearly evident.

Part B is devoted to a succinct statement of a design philosophy and an exhaustive practical example to ilustrate the fine detail. The example is a multiple processor implementation of a protocol for packet switching. It serves to illustrate all aspects of the design alternatives as well as being of wide topical interest.

Part C is a compilation of advanced issues at the system design and implementation levels. Chapter 9 is a condensation and extension of all the previous material. Perhaps it contains the message — the preceding portions of the book supply the background to understand it.

Assumed of the reader is a level of maturity in computer systems which can be gained only by practical experience. It seems necessary to have experience in interrupt-level programming and device interfacing to ensure a practical understanding not only of how to control processors and devices at the assembly language level but also of the practical problems of concurrency. We believe that "hands-on" experience is crucial in this area. Also necessary is familiarity with standard software development techniques using editors, compilers, linkers, loaders and debuggers. Finally, a familiarity with the structure of at least one high level programming language is assumed.

In this book, all programming examples will be given at three levels. PASCAL will be used as a pseudo code to describe logical structure. PL/M will be used where more pragmatic details are considered essential. This approach does not imply that PL/M is unsuitable for describing logical structures or that Pascal is unsuitable for details. The motivation for choosing Pascal for the logical level is its elegance and wide availability. The motivation for choosing PL/M for the pragmatic details is the ready availability of working examples in the authors' laboratory. Finally some INTEL 8080 assembly language code is used in specific examples. However, specific knowledge of these languages is not necessary to follow the material, provided the reader has the maturity provided by the experience recommended above.

High level techniques for describing and programming concurrent systems form an important component of this book. The necessary background material may be provided in a course or two on operating systems. However, such courses are often missing from the backgrounds of the individuals this book is aimed at. Accordingly, no knowledge of such techniques is assumed. Therefore readers with Computer Science or other training which provides this background may find they can skip Chapter 1 and skim Chapters 2, 3 and 5. Others will find these chapters must be read rather carefully as they provide not only the concurrent systems background but also its connection to the subject of multiple microprocessors, all in rather condensed form.

The subject of the microprocessor chips themselves forms a very small part of the material of this book. From the logical point of view, it matters little whether the processors are microcomputers, minicomputers or even maxicomputers. There are, to be sure, certain limitations and characteristics of microprocessors which must be taken into account and these are covered. However, the significant feature of microprocessors is their extremely high performance/cost ratio which replaces the old rules about economies of scale with new ones about economies of specialization. Formerly, economies of scale dictated that as many functions as possible be placed on one processor. It now makes sense to apply multiple processors in a wide range of new applications which hitherto were economically unfeasible, thereby confronting a whole new body of designers with the problems of multiprocessor design.

Finally, we note that the material in this book is aimed primarily at the design of special purpose systems with functionally specialized processors. The construction of powerful, general purpose big processors from collections of small ones may or may not be a viable goal. However, it forms no part of the goals of this book.

Overall the material constitutes a mixture of experience and knowledge. It represents the background we expect our senior designers to have. Some of it may be controversial; all of it is evolving in time. Even so, it is a strong starting position for whatever the future brings.

# Acknowledgements

We are heavily indebted to many people who in one way or another contributed to this book. First of all we are indebted to our academic colleagues and co-workers in the Microprocessor Systems Development Laboratory who not only tested our ideas in practice but also contributed their own ideas to advance the state of the art. And many students became involved in the process of sharpening our focus on key explanations by enduring and criticising early versions of the book.

Jean-Louis Paquet, Walter Brown, and Wayne Redman, first as graduate students and later as co-workers, together with Dennis MacKinnon and Bill Sullivan, forged our first tangible evidence of multiprocessor credibility in the Microprocessor Systems Development Laboratory at Carleton. We put our money where our theses were. And won.

Our academic colleagues Jim Cavers and Bill Bezanson contributed to the maturing of our approach in practice. In particular, Jim Cavers contributed greatly by bending his formidable powers to the problem of designing software for communications protocols. And by producing elegantly structured working designs which form the basis of a number of the examples in this book.

Several graduate students became involved in research activities motivated by and supportive of the Laboratory activities. Mahmoud Sultan contributed to the understanding of the characterization of multiple processor structures, Sikandar Iqbal wrote the first version of Multi Pascal, Andy Yeun provided examples of concurrent systems in Multi Pascal, M. H. Hui wrote a PDP-11 kernel to support concurrent programming, and Bill Robertson provided some of the protocol examples. Others undoubtedly contributed to our education; our gratitude to them all.

The specific contributions of our co-workers in the laboratory are acknowledged in the text.

An original version of the manuscript was read and criticized in detail by William Brown and Dennis MacKinnon. Their suggestions led to several months of agonizing rewriting. Some of it, fortunately, on the shores and in the middle of Lake Ontario.

Finally, as always, there is the miracle of transforming illegible script into the printed word. Our sincere appreciation and thanks to Elaine Carlyle who, with her magic AES word processor, performed this miracle over and over again.

# A DESCRIPTION OF
# CONCURRENT SYSTEMS

This part of the book has one main function: to present a perspective of the structure of multiple processor systems. This perspective exposes the concepts which are used to describe the software and the hardware of such systems. It has the additional virtue of providing the vehicle by means of which these systems can be designed and implemented.

Chapter 1 serves as the interface to the reader. It establishes an overview of the design process and a preliminary description of the three logical components used to describe concurrent systems, in addition to the hardware. Chapters 2, 3, 4, and 5 describe the components and integrate them into a perspective.

The critical theme is simple: in a system in which many activities are occurring concurrently, an hierarchy of critical operating regions is created in which the interaction of the components can be strictly controlled. Understanding and creating concurrent systems becomes a matter of designating and designing these critical regions. The logical components of concurrent systems and indeed the hardware present their own special problems but in isolation these solutions are extensions of what is well known.

Part B will be devoted to deriving a set of design principles for both designing the logical components and their interaction. An extensive example will be used to integrate these principles into a design procedure.

Chapter **1**

# Classical Concepts and Concurrent Systems

## 1.0  Introduction

There has been since the advent of EDVAC and ENIAC a constant activity concerned with improving the performance of computing machinery. In the commercial marketplace, performance has been measured by benchwork programs which are alleged to yield typical run times for a class of problems. Armed with such performance measurements and with systems costs, wise decisions could be made on the purchase or rental of competing processors. Of the many innovations which enhanced performance, technology played an important role. It is clearly possible at this time to re-implement the architecture of earlier computers with modern components, and to find orders of magnitude increases in performance. However, many innovations in architecture also occurred during this time which further enhanced performance. It is with some of these that we are concerned here.

Early attempts to lower the costs of computing installations consisted mostly of keeping them busy. Batch processing was therefore a means of insuring a low cost per minute. And operating systems evolved to allow such use. As the class of users extended to those who were creating their own programs, batch processing became an intolerable bottleneck. The result was the development of multiple access computers or time sharing.

In retrospect the creation of an operating system for time-shared use of a machine seems very simple. Experience had been gained at this time in dealing with interrupts. An interrupt caused suspension of the currently running program, and an interrupt service routine (ISR) saved the state of the processor in a stack (a last-in first-out queue). The interrupt was then serviced; the machine state restored from the stack and the program continued to run. The interrupt activity was totally transparent.

For time-sharing a periodic clock interrupted the processor. On this interrupt the current user state was stored, the user placed in a round robin queue; the next user was selected from the queue, his state installed in the machine and execution resumed until the next interrupt. These time slices allowed many users apparently simultaneous use of a single computer.

3

A user could be viewed at any time as being in one of two states, either running or in a ready-to-run (RTR) queue as shown in Figure 1.1. In order to effect this state transition an operating system maintained the data and manipulated the queues as shown in Figure 1.2.

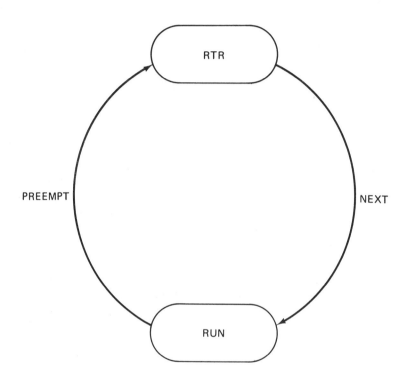

*Figure 1.1: User Operational States in a Time Shared System*

The periodic switching of users was pure overhead and many schemes were introduced to minimize this. Conceptually, however, all schemes accomplished the same results. Provided each user could be treated as a separate entity with no interaction, the operating system remained relatively simple. A problem arises however if a system resource (hard or soft) is being used when run time expires.

A resource (such as a line printer) which is used over several time slots by one user can not be used by another during this same period. The demand for usage by two or more users would appear simultaneous or "virtually concurrent". Techniques were evolved to permit orderly access to such resources.

```
PROCEDURE   PREEMPT;

    BEGIN
        STORE STATE OF CURRENT USER PROGRAM
        ATTACH USER POINTER TO RTR QUEUE

    END

PROCEDURE NEXT;


    BEGIN
        DETACH NEXT USER FROM RTR QUEUE
        FETCH STATE OF NEXT USER AND
        INSTALL ON PROCESSOR
    END
```

*Figure 1.2: Simplified Operating System Procedures for State Transition*

The use of computers as real-time controllers also gave rise to problems of concurrency which were solved by similar techniques or extensions to them.

In some high performance applications, a solution to the requirement for enhanced throughput was to utilize more than one processor. The result was a sharing of the processing, required to execute an algorithm, among the processors. A variety of such multiple processor systems were proposed and built. They were often referred to as super computers to distinguish them from conventional computers. In such configurations, systems resources were shared from logical or from economic necessity. The resultant concurrency problems were more complex than virtual concurrency on a single processor. Earlier concepts were found useful, although a larger set of issues had now to be resolved.

Multiple microprocessor systems inherited all of these classical problems as well as introducing a new set. Microprocessors have inherent performance limitations which are often difficult to accommodate in multiple processor environments. On the other hand their low cost encouraged the development of new configurations in applications which would have been uneconomical with conventional processors.

This chapter reviews the solutions to concurrency in a multiple processor environment. In general a hardware environment is assumed in which one or more processors (each of which may be time-sliced) cooperate in fulfilling the functions of the system. Because of the wide variety of possible hardware architectures, a logical view of the software is required which is as independent as possible of the implementation details. Such a view is forthcoming.

## 1.1    Preliminary Design Concepts

### 1.1.0    Introduction

It often appears that the world may be partitioned into two groups: those who tell others how to design systems and those who actually do it. Those who advise are usually uninhibited, while those who do, often find it difficult to describe what they do. A working system is the manifestation of the designers' knowledge and art and it is with a class of such systems that we are concerned.

A preliminary mutual understanding of the design process seems imperative if Part A of this book is to remain in perspective. Such a view will be presented in this section. It will serve to delimit the range of our discussions and will be refined in Chapters 6 and 7 to a more exact proposal for creating concurrent systems.

### 1.1.1    Systems Design Issues

The overall problem of systems design can be succintly compressed into three inter-related activities:

— Map (i.e., describe) the user requirements (specifications) to a form which captures their essence and creates a set of possible candidate systems.

— Map (i.e. partition) the description in a manner which exposes its logical concurrency.

— Map the partition onto optimal (hardware and software) architectures.

These three mappings are logically related and each mapping requires a backward validation to ensure that it still represents the system. The mappings and their implied extensions will be discussed further in succeeding sections. We note before proceeding that many designs proceed through these three phases with no recognition of their importance or even their existence as logical steps. Often (but not always) this is successful because the choice of the final architecture is constrained. In a real sense the systems designer attempts to tune a predefined architecture to the application. In a multiple processor system, the candidate architectures are numerous and a structured approach is critical if sub-optimal or nonviable systems are to be avoided.

### 1.1.2    Systems Design and Implementation

A perspective of the system designers' tasks is shown in Figure 1.3. We begin by noting that a system is specified by a set of parameters which provide measures of performance which may be tested against the final implementation. It is axiomatic that regardless of the extent to which these parameters are defined and documented, they

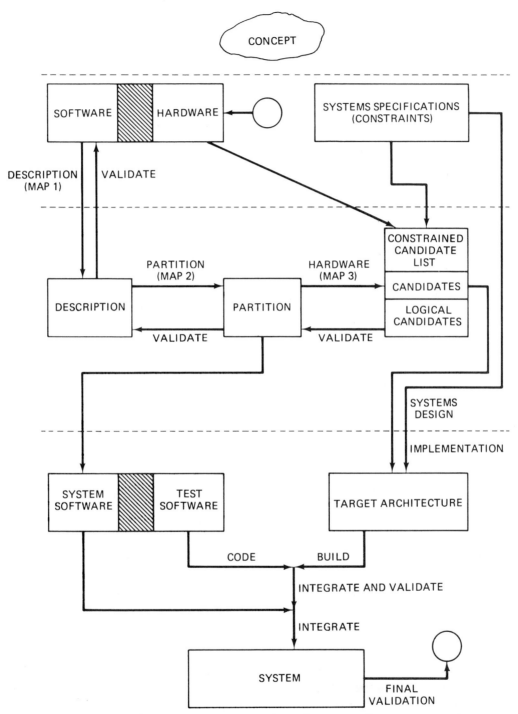

*Figure 1.3: The Design from Concept to System*

are usually not a complete representation of the system. The final system acceptance will certainly have to demonstrate performance in compliance with these measures; however elegant designs and customer satisfaction often result from the intangibles of a well conceived system beyond the formal parameters.

The specifications may also contain constraints which seriously influence the final logical structure of the system (and hence its implementation). These constraints are shown in the broad categories of software and hardware.

Three major steps are required before implementation begins. First, the system (with constraints) must be described in a manner which captures its major features and permits onward manipulation and choices. This step is a modelling step to some extent and the validity of its performance should be confirmed before proceeding. This step is often ignored in systems design on the basis that the specifications are an adequate description in their own right. It is proposed here that this is seldom the case and that a description more suitable to the design process is an essential feature. A description of the logical components of a concurrent system will be discussed in Chapter 2 and 3.

Second, the description must be partitioned to expose those features which facilitate the choice of appropriate hardware. Such a partition should exhibit parallelisms if they exist as well as the precedence relationships of various functions and tasks. Usually several such partitions are possible and should be explored. Validation of each partition with the systems description to insure continuity is essential.

Finally the partitioned description is mapped onto a set of candidate architectures. As shown in Figure 1.3, the list of candidates may have been constrained by some of the original specifications. When the candidates have been selected and validated, a choice must be made and the systems design phase ends. Aspects of hardware will be discussed in Chapter 4.

During implementation the partitioned description corresponding to the chosen hardware is used to define not only the systems software but also the test software. Each implementation activity should be closely related and synchronized.

In the next four subsections we shall consider each step in more detail.

### 1.1.3   Systems Description

The description of the system in a format which is compatible with the ongoing design and which permits a vali- dation with the expressed system performance parameters is an essential first step in the process. To the system designers this becomes the system they are attempting to create. The description should be compatible with their metaphors and techniques. Descriptions in pseudo languages, graphs, tables and charts are all available. In this book we shall propose a combination of graph and pseudo-language techniques to describe systems. Each has a special function and applicability to various phases of the design. After such a description has been agreed upon, and validated by whatever means, the design process can proceed.

### 1.1.4   Systems Partitioning

Systems partitioning refers to the logical division of the description of the system into disjoint sets (of functions and data) which when logically connected together represent the original. The chosen partition plays an important role in determining the characteristics and performance of the final product. Good partitions lead to modular systems with well defined functions and minimal interprocessor communications requirements. Thus the software tends to have these same characteristics. Good partitions lead to good systems; however, very little is known about such techniques. There are heuristics which seem to work and often partitions are intuitively obvious. An *access graph* approach will be introduced and used throughout the book. This technique tends to expose good partitions, at least for moderate sized systems.

A partitioning of functions is usually accompanied by the requirement to consider the partitioning of data. Data tables tend to fall into three categories: First, they are uniquely associated with a function and hence are partitioned with it. Secondly, they may be shared but static and could therefore be duplicated. Thirdly, they can be shared but dynamic (i.e. alterable). In this case precautions are necessary to prevent unsynchronized alterations by different functions. Indeed the constraints imposed by data and the flow of data through the system tends to drive the overall logical design. An example will be discussed in Chapter 7.

### 1.1.5   Hardware Mapping

In an open ended design the hardware architecture which executes the partitioned system depends on several performance specifications of the system and features of the partition. A major purpose of this book is to expose the alternatives in such a way that an optimal choice can be made. If the design is not open ended, that is, constraints exist on the choice of the processor configuration, then fine tuning will be based on the same alternatives. Iterations may be required to alter the original partition to accommodate the hardware. Indeed it may prove impossible to satisfy the specifications unless the hardware constraints are modified. Clearly the suggested modifications require a knowledge of all the alternatives.

### 1.1.6   Systems Implementation

Implementation issues will be treated in detail in Chapters 7 and 10. For completeness we observe here that the chosen partition and the hardware specify the organization of the system software. An important task is to devise test software which can be used to test the hardware and to permit integration of the systems software and the hardware. This often overlooked activity can substantially reduce the final integration (and debugging) phase. The first step is of course the validation of the prototype with the system parameters. The test software can often be designed to serve, in part at least, this function. The constant attention to interim validation at each step of the systems design tends to increase the probability that this step is routine.

## 1.2   Concepts and Components of Concurrent Operating Systems

### 1.2.0   Preliminary Comments

The term "operating system" has a wide variety of meanings. Throughout this text the term will be used to refer to any system of programs which communicate with each other and with the external world. Specifically it is not necessary to infer the provision of services to "user programs". Indeed, in special purpose, non user-programmable systems, the operating system may be the union of all the software.

In general, operating systems perform functions such as initiating external activities, responding to external events, coordinating internal activities, scheduling work and allocating resources. In our view, these functions may be performed in a centralized or distributed fashion. There is no implication that the operating system is necessarily a single, monolithic "monitor", as is often the case in general-purpose, user-programmable systems. Indeed, as we shall see, there may be many "monitors" in our operating systems, to provide different, special services to the concurrent programs which form the operating system.

In the following sections key features and concepts from concurrent operating systems design will be grouped, discussed and biased toward multiple processor implementation. The first concept is a *Process*.

### 1.2.1   A Process

The concept of a *process* is fundamental to the description of a multiple processor operating system. Concurrent systems are too complicated to think about in general without breaking them into familiar building blocks. The highest level familiar building block appropriate for concurrent systems is the sequential program. If a concurrent system can be described, designed and developed as a set of (possibly interacting) sequential programs, then we may have some hope of keeping its complexity under control. Because concurrency may imply sharing processing resources, a sequential program must be augmented by a named set of state information which allows the program to be restarted at any point. This generalization of a sequential program is called a "process".

A process is thus an abstraction which allows a perspective of systems functions at a level above the program code. Two processes could, for example, share common subroutines or procedures and yet be viewed as separate entities while simultaneously executing the same code on different data. Processes are logical entities which do things in the system, e.g., the command interpreter process, the packet reception process, etc. They are the primary active elements of concurrent software. The perception of processes will be sharpened as we proceed.

Processes are grouped into two types:

1) Independent processes: neither compete nor cooperate with each other.
2) Concurrent processes: interact with each other. They may cooperate, share and compete for resources.

Concurrent processes imply some form of cooperation. It is with the implications of this that the remainder of this section is concerned.

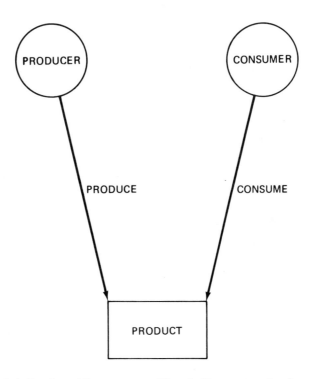

*Figure 1.4: Producer/Consumer — Simple Representation by an Access Graph*

## 1.2.2   The Producer/Consumer Example — Case 1

A classical problem usually referred to as the producer/consumer problem will be used several times to illustrate both problems of and solutions to concurrency. Here we are concerned with stating the problem and exposing the most obvious issues which must be resolved.

A producer process is assumed which creates an output to be utilized by a consumer process. For example, the producer might accumulate a buffer of data from an input device which is to be processed by a consumer process. To depict the situation graphically we draw an access graph, as shown in Figure 1.4. An access graph is a kind of software "wiring diagram" for concurrent systems, in which the active components (in this case, processes) are shown as circles and the passive ones (resources) as rectangles. The arrows indicate access rights, or "connections". For the situation of Figure 1.4 to be interesting, it is assumed that some continuing cooperation is required between the two processes to coordinate the transfer of full buffers and the return of empty ones, or to insure data integrity, etc. Four major problems can be identified if orderly cooperation between the processes is to be realized: *mutual exclusion, synchronization, deadlock, determinancy.*

*Mutual Exclusion*:   This refers to the necessity of enforcing the strict sequential use of a resource by competing processes until the task is finished. With reference to Figure 1.4, mutual exclusion is required between production and consumption of the product. Several mechanisms are available for ensuring this. They will be discussed in Section 1.2.3.

*Synchronization*:   Cooperating processes may require some form of synchronization in order to coordinate a sequence of operations on shared resources. With reference to Figure 1.4, synchronization is required with respect to the availability of products and the return of used ones. Synchronization is a major requirement in multiple processor systems and two important techniques will be discussed in detail in Section 1.2.3 which lead to the concept of a monitor discussed in Section 1.3.2.

*Deadlock*:   This term describes a situation which occurs when processes are mutually waiting for events caused by other processes. With reference to Figure 1.4, deadlock may occur if the producer/consumer relationship is bilateral. Then each process may be waiting for a product to be produced by the other; while waiting, nothing is produced by either process and a deadlock exists. In multiple processor systems with concurrent processes, mysterious forms of deadlock can sometimes occur. In very complex systems it is difficult to guarantee that deadlock will not occur. However, three conditions have been isolated which will lead to deadlock:

i)   each process can claim exclusive control over the resources it holds

ii)   each resource assignment is non-preemptive

iii) there is a circular wait; a process holds a resource while waiting for another
to complete its task

Generally, a system is designed to prevent deadlock by careful attention to these three factors. In larger systems the operating system can attempt to avoid deadlock by analysing the current state of the system or it can detect deadlock and institute recovery procedures.

Further discussion of deadlock problems will be left until the design example in Chapter 7.

*Determinacy*:  This problem refers to the possibility that processes which can concurrently access common data could alter this data by an interleaved sequence of operations which invalidates it. Special precautions are necessary to ensure that alterations to common data are coordinated in such a way that the data is always left *determinate*. Techniques will be discussed in Chapter 6 for insuring determinancy. Solutions to this problem require a combination of techniques used for the exclusion and synchronization of processes. It will be treated in more detail in Chapter 3.

## 1.2.3  Critical Regions — Semaphores

The key to the orderly interaction of concurrent processes is the concept of a *critical region* of code. Processes are constrained to enter these regions according to some pre-established discipline. It is in these regions of code that strict sequentiality can be enforced and the human intellect can cope with concurrency.

*A Critical Region* is a section of code in a concurrent system which is controlled so that only one process at a time can enter.

Critical regions of code are analogous to regions of railway track in which only one train at a time can enter. Entrance and exit to these regions is controlled by a semaphore. A train approaching a critical section observes the semaphore, if it is UP, it indicates another train is in the section and the approaching train must WAIT. When the semaphore drops the waiting train can proceed. Concurrent systems use this semaphore concept to control access to critical regions.

Consider first that a Boolean variable exists called FLAG. This variable can be a location in memory (i.e., a software flag) or a flip-flop (i.e., a hardware flag). The implementation details are not of concern at this time. Assume that if FLAG = TRUE, then the critical region is being used and vice versa.

A process wishing to enter a critical region examines the FLAG. If it is TRUE, the process must wait; otherwise it sets the FLAG = TRUE and enters the critical region. Upon completion it resets FLAG.

A logically simple construct; however on reflection several problems emerge:

1)  What happens if two processes simultaneously attempt to examine the FLAG?
2)  How does a process wait if the FLAG = TRUE?
3)  How is a waiting process restarted when the FLAG is reset?

The first problem is the most difficult. Clearly if asynchronous requests are possible then simultaneous requests are always a possibility. A complete analogy with asynchronous interrupts is apparent and some form of hardware arbitration is finally necessary which allocates priorities according to some algorithm. For now, some form of arbitration is assumed to exist (arbitors will be discussed in Chapter 5). Thus manipulations of the FLAG are assumed arbitrated and indivisible:

An *indivisible operation* (including sequences of code) cannot be interrupted by any feature of the system (either software or hardware).

An indivisible operation is a short critical region. The mechanisms for creating indivisible primitive operations depend on many factors and examples will be explored in Chapter 5.

Suppose that the flag is examined and manipulated by an indivisible procedure as follows:

```
LOOP: [IF FLAG = FALSE THEN FLAG := TRUE
       ELSE] GO TO LOOP;
```

The brackets [] indicate indivisibility. Any process wishing to enter a critical region performs this test. It loops on this piece of code until it finds the FLAG = FALSE. Note that while the operation is indivisible that other processes could gain access to the flag outside the brackets. To enter and exit a region of critical code a process would call as follows:

```
PROCEDURE:  ENTER CRITICAL REGION;
       BEGIN
          LOOP: [IF FLAG = FALSE THEN FLAG := TRUE
                 ELSE] GO TO LOOP;
       END;

          (CRITICAL REGION)

PROCEDURE:  EXIT CRITICAL REGION
          BEGIN
           [FLAG := FALSE]
          END
```

We note in passing that the loop procedure is equivalent to a *test and set* (TAS) instruction which is supported by newer microprocessors (including indivisibility). In Chapter 4 a hardware implementation is shown which accomplishes the same result.

The obvious disadvantage of the LOOP procedure for testing a flag occurs when the flag is TRUE. In this case a process continues to execute the procedure until the flag is set FALSE by another process (the one already in the region). This form of waiting is referred to as a *busy wait*. This busy waiting is essentially nonproductive and in some configurations it can slow the whole system.

To avoid the overhead of *busy-waits* a process could be suspended if a critical region is occupied, and a new process allowed to run on the processor. This new process could be selected from the RTR queue as though the time-slice interrupt had occurred. Clearly this should enhance the productive instruction throughput of the overall system. The suspended process could be put in a separate queue to await the availability of the critical region. This separate queue creates a new process state called a WAIT state as shown in Figure 1.5. To accomplish all this requires a more complex entry procedure than the simple loop. The new concept is called a semaphore.

A *SEMAPHORE* is a logical construct which may be used to control entry to (and exit from) a critical region without busy waiting. It includes the appropriate queues and procedures for suspending a process.

In its simplest form, a semaphore is composed of a queue and a non-negative counter. The counter is an extension of the Boolean flag of the LOOP construct. It records the number of processes which may pass the semaphore without being suspended on executing a wait. When used for critical region protection this number is initially one and drops to zero when the critical region is busy. The counter may also be used to record the number of processes which have been suspended (and hence which must eventually be restarted) by allowing it to take on negative values.

Operations on a semaphore are defined as follows:

```
TYPE SEMAPHORE = RECORD S : COUNT;   INTEGER
                               QUEUE : PROCESS-QUEUE
                       END;

PROCEDURE SEMAPHORE-INIT (SEM : SEMAPHORE);
    BEGIN
        SEM.COUNT:=SO;   SO=1 FOR CRITICAL REGION PROTECTION

        INITIALIZE SEM.QUEUE (DEPENDS ON QUEUE DISCIPLINE)
    END;

PROCEDURE WAIT (SEM:SEMAPHORE);

    BEGIN
        [SEM.COUNT := SEM.COUNT-1;
        IF SEM.COUNT  < 0THEN BEGIN
            IDENTIFY CALLING PROCESS;
            ENTER IT IN SEM.QUEUE; (I.E., SUSPEND IT)
            START NEXT PROCESS IN READY-TO-

        END]
    END;
```

```
PROCEDURE SIGNAL(SEM:SEMAPHORE);
    BEGIN
        [SEM.COUNT := SEM.COUNT + 1;
        IF SEM.COUNT ≤ 0 THEN BEGIN
            REMOVE NEXT PROCESS FROM SEM.QUEUE;
            PUT IT ON READY-TO-RUN QUEUE;
        END]
    END;
```

Where again [] indicates indivisibility. The WAIT state in Figure 1.5 is used to indicate a process is in the associated semaphore queue. Each system semaphore has a corresponding queue and a process in any of these queues is considered to be in the WAIT state.

Thus:

To use a semaphore for critical region protection its counter is initialized to one. If a process wishes to enter a critical region:

— it performs a WAIT;
    if the counter equals one, it proceeds into the critical region; if the counter is less than one, it is suspended in the WAIT queue and the next process in the RTR queue is allocated to the processor.

When a process leaves a critical region

— it performs a SIGNAL;
    if the counter is negative, the next waiting process is moved to the RTR queue.
A semaphore has obvious overhead associated with its implementation which implies both execution time and storage space. Efficient implementations are therefore of importance and will be dealt with later.

Finally it is noted that operations on the semaphore are also critical regions. Each operation is a longer critical region than the Loop procedure. It could in fact be implemented by the ENTER procedure using the Loop discussed earlier.

Clearly a trade-off begins to emerge. If a Loop (which is a critical region) is used to create a WAIT (which is a critical region) which is used to create the desired critical region, is there a net gain? The answer lies in the relative lengths of time required for each region. The desired critical region is assumed to be longer than the WAIT.

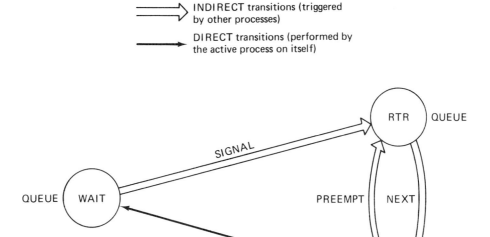

LEGEND

⟹ INDIRECT transitions (triggered by other processes)

→ DIRECT transitions (performed by the active process on itself)

*Figure 1.5: Process State Diagram with WAIT States*

Thus:

— The Loop requires only a few assembly language instructions, but continues to execute until entry is gained

— The Wait is relatively short (compared to the critical region being created) and quickly suspends the active process, allowing other work to continue.

The net result must be an increase in throughput. In examples in the following chapters we will use both mechanisms to establish critical regions.

### 1.2.4   Interprocess Communciation

The producer/consumer example referred to in Section 1.2.2 can now be re-examined. A critical region of code can be created using semaphores. In this region the common data can be manipulated. For example, in a simple case a pointer to a full data buffer could be passed to the consumer process. And eventually the pointers to empty buffers returned. This procedure solves the exclusivity problem first mentioned in Section 1.2.2. Semaphores can also be used to synchronize the producer and consumer with respect to the events "all buffers empty" and "all buffers full".

### 1.2.5   Process Scheduling

A critical region can be created, as discussed in the last section, in which the exclusive use of common resources can be enforced. And processes can also synchronize their actions with respect to events by using semaphores. However, in complex problems with many events, programs become tricky; unstructured use of semaphores is like unstructured use of the "Go To" statement in this respect.

A mechanism is required within the critical region which can schedule process interaction in a more structured way. This can be accomplished by associating status information with the critical region and by including internal mechanisms for blocking a process if the status information indicates that certain required conditions are not valid. Blocking a process inside a critical region presents the same logical problem as creating the original critical region. That is, the process cannot execute a sequence of code because a status condition is not valid.

BLOCK thus consists of a WAIT on a condition recorded in a status variable. If the condition is not valid, the process is suspended in a queue associated with the condition. When this happens the semaphore protecting the critical region must be signalled to allow the entry of another process. The BLOCK state is shown in Figure 1.6. Release from the block state is performed by the process which satisfies the status condition. Clearly this scheduling is a separate problem and the design of the mechanisms become most important.

A *condition variable* is a combination of a status condition and queue. Condition variables may be implemented directly or by using semaphores private to the semaphore.

The critical region has now assumed a fair responsibility for correct system operation and should be viewed as a separate logical entity. It will be called a *Monitor* and discussed in detail in Chapter 3.

A *MONITOR* is a generalized critical region in which interprocess communication and scheduling is effected using condition variables.

### 1.2.6   Resource Manipulation — Monitors/ Kernels

Thus far we have proposed that common resources could be shared by processes by creating a critical region of code. In this region, buffers, queues or other data structures

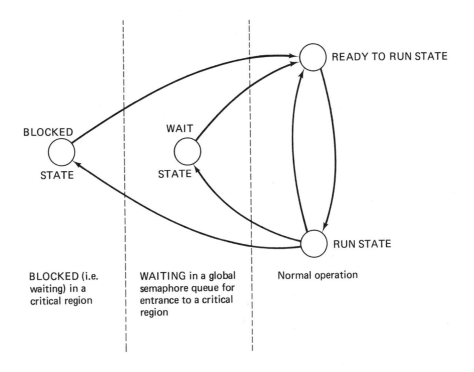

*Figure 1.6: Blocked and Wait States*

could be manipulated. As well, scheduling of processes dependent on certain status conditions can be arranged. The code for this critical region could become complex and the scheduling mechanisms varied. The whole region can be considered as a separate logical component and was referred to as a Monitor.

If suitable calling conventions are maintained, the monitor code can be designed, implemented and debugged separately. The processes using the code are said to be operating "in the monitor", while executing it. Chapter 3 will be devoted to describing the organization and use of monitors. In Chapter 7 an example of the creation of monitors as a natural part of the logical design of a system will be given.

In a large concurrent system it is now clear that the processes state graph could contain many queues in addition to the ready-to-run queue. Indeed much of the overhead of such systems is the moving of processes through the state graph (i.e., from queue to queue).

The major state transition might be considered to occur between the RTR queue and the RUN state. This involves the maintenance of process state information and the queue maintenance procedures. The WAIT and BLOCK transitions also involve moves requiring access to these data structures. It seems reasonable to concentrate all process state transitions in a single logical entity. It will be called the kernel.

The *KERNEL* is a critical region responsible for manipulating the common data structures associated with state transitions.

The kernel is a critical region of code like a monitor. Logically it is the fundamental monitor of the system. Any activity in which a process state may change will involve the kernel. In this way an orderly transition of process state can be enforced. Kernels will be discussed in some detail in Chapter 5.

Other functions can be allocated to the kernel. In a multiple processor environment the kernel must maintain data structures which, for example, identify processors which will run only certain processes. Indeed the kernel can also be used to provide isolation of the hardware from the logical structure of the system. In this respect the kernel transforms the hardware into a "virtual machine" on which the concurrent system runs. This view will be projected in Section 1.3 and amplified in the remainder of the text.

One implication of the tasks allocated to the kernel is already evident. If the kernel is to maintain process state information this implies that WAIT and SIGNAL (as well as BLOCK and UNBLOCK) must be performed by the kernel. Indeed since the kernel itself is a critical region the semaphore operations are also a critical region (as part of the larger one). If the kernel were sufficiently short then access could be gained by a loop procedure. And after entry the appropriate semaphore operation called. This implies parameter passing (and possibly returning) which further implies some formal protocol for kernel calls. These matters will be explored in Chapter 5.

### 1.2.7  Hardware Considerations

The logical structure of a concurrent system must eventually be reduced to software and partitioned to run on the available processors. Obviously the partition will be affected by the hardware architecture. Clearly it is desirable to create such architectures as will permit the widest possible variations in logical complexity to be easily mapped. This is desirable but not always possible. A homogeneous architecture is desirable in the sense that the system can be mapped without hardware constraints. The importance of this mapping problem requires a wide knowledge of hardware structures. And this will be covered in Chapter 4.

The design approach suggested in Chapter 6 suggests that the kernel design is squeezed between the hardware constraints and the logical requirements of the system. The kernel designer and the hardware designers must obviously be knowledgeable of each other's problems.

## 1.3  A Description of Concurrent Systems

### 1.3.0  Some Guidelines

The previous section was devoted to a range of basic concepts which present a perspective of concurrent systems. The next four chapters will examine these concepts in more detail. These chapters will provide not only a means of examining and discussing the logical structure of concurrency but, with the addition of some design insights, a total algorithm for creating such systems.

The next four chapters are interrelated and do not stand in isolation; therefore, in this section, an overview will be presented so that a point of reference can be maintained until all components are in place. The description of the logical structure is based on a combination of classical concepts modified to fit the multiple microprocessor world. It is appropriate here to introduce the practical requirements for such a description. The overall driving force is to design and implement, within a budget, systems which can be maintained over their lifetime.

The design process was discussed briefly in Section 1.1 and it will be expanded in Chapter 7. Concurrency introduces special problems in description, implementation and debugging. It is necessary to provide a logical description which provides for not only the design at the systems level but implementation and debugging as well. In addition the final logical structure must run on a multiple processor architecture. This implies a partition of the logical description to accommodate various target architectures. The conceptual basis must therefore be capable of supporting these requirements. The systems design must be software engineering at its finest to avoid catastrope. In particular modularity and transparency are important principles. Functions must be implemented in modules. Interfaces across modules must be carefully designed. Modules and their interfaces must be easily understood. And to the greatest extent possible, the complexity of scheduling and processor architecture, should remain transparent. As much as possible, process code should be written as though for a dedicated uniprocessor.

The relationship of processes to monitors is often shown on an *Access graph*. An example of such a graph is shown in Figure 1.7. The arrows denote access rights of processes to monitors. Note that the kernel is not shown. This is consistent with the view of a kernel as a virtual machine which provides the services to implement the graph.

Logical modularity between all the components of our description is achieved as shown in the onionskin view contained in Figure 1.8. This view, with an accompanying access graph, describes the logical structure of the system. Design principles for multiple processors allow the graph to be partitioned and mapped onto hardware (as explained in Chapters 6 and 7).

By relating the concepts introduced in previous sections, a finer perspective of concurrent systems can now be obtained. This is the perspective shared by the authors and it permeates and is implicit in all that follows. The "onionskin" representation of Figure 1.8 contains the various levels of complexity of the system from the processes to the hardware. Looking at the outside we call the whole a Concurrent High Level Language (CHLL) machine. A summary of the components of the onionskin follows:

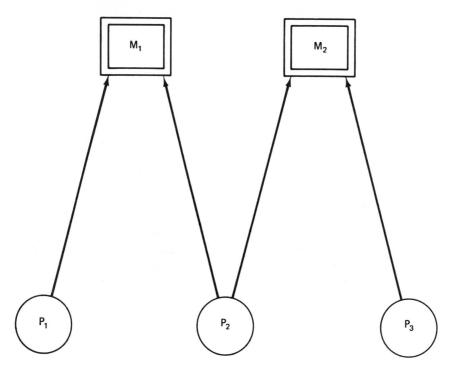

*Figure 1.7: Access Graph — Example*

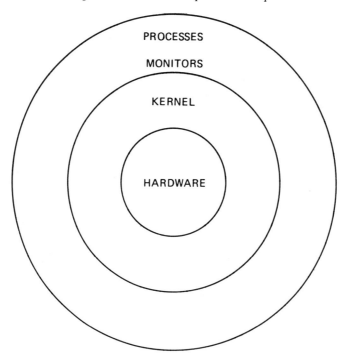

*Figure 1.8: System Logical Model — Onionskin*

### 1.3.1 Processes

The outermost layer contains the processes which perform the logical tasks required to define the performance of the system. These tasks may be sequential or concurrent. Processes are the active components of the system. In a well designed system, resources are manipulated by procedures which render concurrency transparent to the programmer. This requires a strict adherence to prescribed calling disciplines. The adherence to these disciplines will guarantee a solution to concurrency problems in a well designed support system (i.e. the inner layers of the onionskin). These disciplines or calling protocols could be enforced by a suitable compiler for a language which supports concurrency. No such compilers are available for microprocessors (at time of writing) although concurrent PASCAL might fulfil this gap in time. As a result the protocols must be enforced by programming conventions adapted to simulate the action of a compiler. With the proper adaptations PL/M, Pascal or even Fortran can be used to write suitable processes. The use of PL/M and Pascal will be discussed in Chapter 8.

The definition of process from system specifications is not an algorithmic activity. "Finding" processes is a creative design activity. A description of process interaction is given in Chapter 2. And a design example illustrating their definition from systems requirements is given in Chapter 7.

### 1.3.2 Monitors

The next layer below the process contains the monitors. A monitor is usually defined each time a common resource is identified. This is not as straightforward as it sounds because such common resources are often software modules which perform common functions which are only identified during the software design activity. Indeed "finding" monitors, as well as "finding" processes, is a creative design activity. Many iterations are usually necessary before an acceptable description is found. The criteria of acceptability are vague; however, at a minimum it must perform the logical functions required by the system specifications. And logical simplicity is a very desirable attribute (even though difficult to define). Monitors are passive components in the system in the sense that they perform functions when called by processes.

Having identified a common resource and created its associated monitor, it is now required to give the monitor its character. What are its operations, its scheduling mechanisms, its data structures, etc.? The description of monitors is left until Chapter 3. Examples of the natural evolution of monitors as part of the system design are shown in Chapter 7.

### 1.3.3 Kernel

The kernel provides for all communication and synchronization between processes and monitors. It also directly interfaces with the processors of the system. Ideally it is the only layer which is cognizant of the processor topology. This is seldom fully achieved, at least in microprocessor systems. The kernel is the fundamental monitor of the system.

The kernel is a critical region which performs different functions embodied in a set of kernel procedures. These procedures will be called PRIMITIVES because of their indivisibility. These primitives characterize a virtual machine composed of the kernel and the hardware. Indeed with a reasonable choice of primitives, a general purpose virtual machine can be designed to implement a wide variety of process/monitor access graphs, depending of course on such factors as response time, etc. The design of kernels will be presented in Chapter 5. The intimacy of the kernel and the hardware suggests an equally close implementation relationship. This is so, and the considerations involved in realizing the kernel functions in an optimal way involve the hardware topology.

### 1.3.4   Hardware

The utilization of microprocessors in multiple processor architectures has been motivated by good price/performance comparisons. They were not designed for this purpose however and are often adaptable only with considerable ingenuity. However, their widespread use will undoubtedly influence future processor designs and a rapid increase in such systems is foreseen. In Chapter 4 a wide ranging discussion of interconnection topologies is presented to lay the groundwork for the anticipated changes in available hardware.

### 1.3.5   CHLL Machines

The various components discussed thus far and their interaction must be described and documented in such a way that implementation is facilitated. Such a description specifies a concurrent high level language machine (CHLL). The features of a CHLL machine can be discussed with the following:

a)  An access graph which connects the following:

   i)  Active components

      — Processes

      — Interrupt Service Routines (ISRs)

   ii) Passive components

      — Monitors

      — Semaphores

      — Subroutines

      — Data Structures

The arrows on the connecting lines in the graph indicate access rights; while active a process has exclusive control of the passive component. Figure 1.9 gives conventions for access graphs to be used in this text.

b) An algorithmic description of the processes showing run-time connection sequence and parameters passed to passive components.

c) A description of the external services provided by each passive component.

d) A description of "connection protocols" between active and passive components by means of which active components acquire and release exclusive control of passive components.

e) An algorithmic description of the operation of each passive component.

The operating software can be written in any concurrent language which supports the connection protocol. In practice this usually means using an available high level language and kernel and writing the connection protocols explicitly.

Connection protocols are the mechanisms by which processes acquire and release control of passive components. These mechanisms should in general

i)   check access rights

ii)  set up the hardware/software connection

iii) acquire and relinquish exclusive control of the passive component

A concurrent high level language would provide these features as part of its structure. If such a language is not available then a "friendly" high level language must be adopted. Such a language should have:

— good control and data structures

— reentrant subroutine calls

— an ability to manipulate code segment labels as addresses

— an ability to interface directly to interrupts and I/O devices

— a modular development capability with linking facilities, including the ability to use assembly language modules

A more detailed discussion of high level language for this application is reserved for Chapter 8.

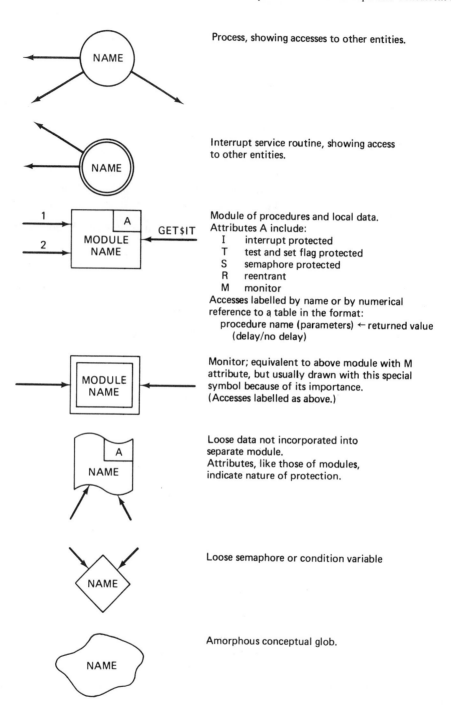

Process, showing accesses to other entities.

Interrupt service routine, showing access to other entities.

Module of procedures and local data.
Attributes A include:
I    interrupt protected
T    test and set flag protected
S    semaphore protected
R    reentrant
M    monitor
Accesses labelled by name or by numerical reference to a table in the format:
procedure name (parameters) ← returned value
(delay/no delay)

Monitor; equivalent to above module with M attribute, but usually drawn with this special symbol because of its importance.
(Accesses labelled as above.)

Loose data not incorporated into separate module.
Attributes, like those of modules, indicate nature of protection.

Loose semaphore or condition variable

Amorphous conceptual glob.

*Figure 1.9: Access Graph Modules*

## 1.4  Summary

This chapter serves as the interface for the reader. The next four chapters serve to put together the logical components required to describe concurrent systems. Part B is concerned with using these components as part of the overall design process. The interface is intended to serve readers with widely ranging backgrounds.

Multiple processor systems imply concurrency. Concurrency must be described in a logical way which can be implemented in software, which furthermore must be mapped onto hardware. A consistent description which facilitates this whole process is being proposed here. It also facilitates system documentation, and subsequent maintenance.

At this point we have established two preliminary perspectives:

1) A design process which motivates the need for an adequate systems description.

2) A set of interacting components which allow a logical description of a concurrent system.

In the next four chapters each component of the logical description will be explored in detail. At the conclusion of Chapter 5 a more comprehensive picture of such systems should emerge.

## 1.5  References

An appropriate paper on deadlock problems is:

E.G. COFFMAN, JR., M. ELPHICK, and A. SHOSHANI, "System Deadlocks", Computing Surveys, Vol. 3, No. 1, June 1971, pps. 67-78.

A good overview of software engineering principles is contained in:

D.T. ROSS, J.B. GOODENOUGH, C.A. IRVINE, "Software Engineering: Process, Principles and Goals", Computer, May 1975, pp 17-27.

Chapter **2**


# Processes — The Active Components


**2.0    Introduction**


In this chapter the first of the four components of a concurrent system will be explored. From Chapter 1 we recall that processes are the active components in an onionskin-like array of the monitors, the kernel and the hardware. Concurrent processes (as any process) consists of sequential code and private data structures. In addition of course some sharing of global system resources with other processes is implied. We are not concerned here with the generation of process code, but with the mechanisms for interprocess synchronization and coordination.

The mechanisms for synchronizing and coordinating concurrent processes can be extended to interrupt service routines (ISRs) if some precautions are observed. Interrupt service routines are degenerate processes which are scheduled by the hardware instead of the kernel.

Section 2.1 provides a brief introduction to processes.

In Section 2.2, intercommunication mechanisms are explored, between processes, and between processes and interrupt service routines. This section illustrates the need for monitors and device handlers. A keyboard operating system is used to illustrate these ideas in Section 2.3, to illustrate the use of access graphs and to raise some design questions.

Section 2.4 discusses the problems of describing, initializing and creating a process. The full impact of this procedure will be better focused after studying the details of the kernel in Chapter 4.

A motivation of this chapter is to show the need for structured, high-level interprocess communication mechanisms by illustrating how unstructured, low-level ones lead to complexity. A further motivation is to illustrate the difficulty of the creative art of "finding" processes; this paves the way for further discussions in Chapter 7.

## 2.1 Specification of a Process

It was pointed out in Chapter 1 that "finding" processes in the original systems specifications is a creative activity. No algorithm can be proposed which would yield a process from the system specifications. There is an art to this creation which depends on experience and is guided by criteria appropriate to a particular environment. In Chapter 7 extensions to the design procedure proposed in Chapter 1 tend to yield processes based on a set of principles appropriate to a multiple processor environment. At this time, only broad guidelines are possible. For example,

1. A process should be logically simple with closely related functions to perform.
2. A process should contribute to the logical simplicity of the overall system.
3. The process code (in a high level language) should be short enough to be easily comprehended.

The thrust here is to create logically simple structures which can be understood, designed and debugged easily. A maturing concept of process creation and definition will occur throughout Chapter 7.

At a lower level, a process must be specified in the data structures of the kernel. This specification is contained in a process control block (PCB) which contains sufficient information to restart the process from any of the various queues in which it may be suspended. A PCB might contain for example

```
Type Process-Control-Block =

    record

        Process Index: (Name of the Process)
        Microprocessor Index: (processor on which it runs)
        Priority: (if applicable)
        Stack pointer: (Private stack)
        Starting address: (initial program counter)
        Register state: (machine registers contents when
        suspended)
```

With this information the kernel can obtain information for state transfers. This concept is extended in Chapter 5.

## 2.2 Interprocess Communication

### 2.2.0 Process Interfaces — Preliminary

A concurrent process interacts in some manner with other processes in the system. Conflicting criteria are often encountered when attempting to specify the intercommunication mechanisms. On the one hand there is the desire to isolate the programmer, who writes process code, from the complexities of concurrency. Indeed, a principle is:

Principle 2.1:

Transparency; create and maintain a software environment in which all concurrency problems are transparent at the process level.

This principle is invoked by most user-programmable, general purpose computer systems. In such systems the operating system is a single monolithic monitor which handles all concurrency problems. On the other hand this approach is inflexible and involves considerable overhead in generalized scheduling and communications mechanisms which can create violations of other specifications (e.g., response time). In situations in which trade-offs must be considered, we propose:

Principle 2.2:

Strategic Retreat; introduce concurrency problems, at the process level, reluctantly.

The strategic retreat principle will reappear in many guises as we proceed. Clearly we are always aiming at a system in which the layers of complexity inherent in the whole system are introduced in a controlled manner which violate as few other design principles as possible.

## 2.2.1   Process to Process — Low Level

In Chapter 1, the producer/consumer example was used to illustrate interprocess communication. In this section this example will be extended in more detail and used to motivate the requirement for monitors.

Consider the scheme illustrated in Figure 2.1. Two semaphores, SEM1 and SEM2, are used to synchronize the use of common buffers with respect to the conditions "all buffers empty" and "all buffers full", respectively. For simplicity, assume a set of fixed size buffers are available. The producer process fills a buffer and attaches the address to QUEUE F with an ATTACH procedure. The consumer process gets a full buffer to work on by a DETACH procedure which yields the next entry in the queue. The details of these procedures are unimportant at this time. However we note that they form a critical region to be protected by a semaphore.

To return an empty buffer the consumer process attaches the empty buffer address to QUEUE E and the producer process detaches empty buffers from this queue as required. The proposed synchronization is as follows:

To pass a full buffer, the producer signals SEM1:

```
                        –
                        –
            ATTACH (QUEUE F, BUF ADDRESS)
            SIGNAL (SEM 1)
                        –
```

The consumer is suspended in SEM1 if all buffers are empty:

```
    -
    -
WAIT (SEM 1)
DETACH (QUEUE F)
    -
```

To return an empty buffer the consumer signals SEM2:

```
    -
    -
    -
ATTACH (QUEUE E, BUF ADDRESS)
SIGNAL (SEM 2)
    -
    -
```

The producer is suspended in SEM2 if all buffers are full:

```
    -
    -
WAIT (SEM 2)
DETACH (QUEUE E)
    -
```

Note that separate full and empty queues are not strictly necessary but that both semaphores are necessary.

The scheme proposed here is certainly simple. Will it work? In order to guarantee that the required synchronization will occur and that no untoward events (such as deadlock) might also occur requires exhaustive argument. In particular it is necessary to show, for example, that no timing constraints are implicitly assumed. It is not proposed here to prove this system works as proposed but to suggest that it clearly violates principle 2.1.

In particular

> Process code is used to manipulate the common resources (ATTACH and DETACH). And scheduling is clearly the responsibility of the process programmer.

Consider for example the problem of introducing more producer and consumer processes; modularity is poor and the result is complex. The prospect of debugging and maintaining a larger system of this kind is frightening. A more desirable system is shown in the next section.

### 2.2.2  Process to Process — High Level

The low level scheme proposed in the previous section suffered from a lack of modularity and an excess of logical complexity. The more desirable approach is to concentrate the scheduling problems in a single critical region of code referred to earlier as a monitor. In this section we can only illustrate this, leaving a detailed discussion for the next chapter.

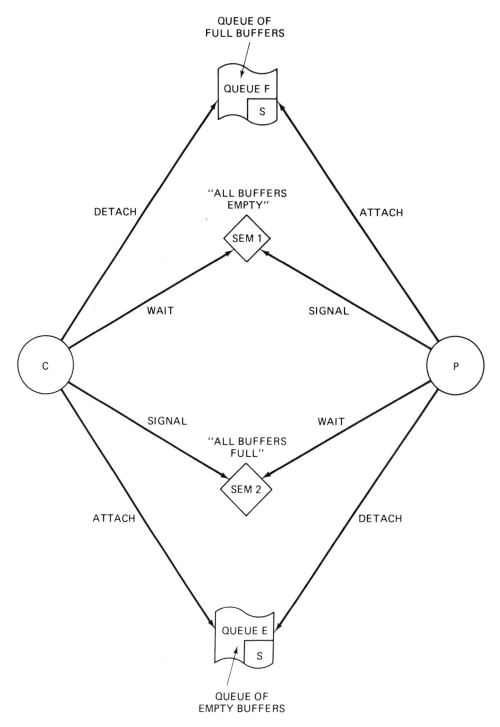

*Figure 2.1: Process Passes a Buffer to Another Process*

In Chapter 1 a monitor was proposed as a critical region of code created by a semaphore. All processes WAIT on this semaphore and once inside the monitor strict sequentiality is guaranteed. The monitor can have as well private condition variables which can be used for scheduling or for any other purposes.

Consider the scheme shown in Figure 2.2. A producer process is to be synchronized with several consumer processes by a monitor.

Logically the system works as follows:

1. Any process can call the monitor (with a set of parameters) by an entry procedure

2. The entry procedure includes a WAIT on the monitor gate

3. On entry the parameters are examined by the monitor and the calling process is serviced or waits on a condition variable until it can be serviced.

The details of performing these actions will be treated in Chapter 3. For now we observe that the process scheduling code can be debugged separately. Processes communicate by procedure calls with rigid parameter passing mechanisms. Aside from the possibility of increased overhead, the system satisfies all the principles enunciated previously.

### 2.2.3   Process to Interrupt Service Routine (ISR)

Interrupt service routines (ISRs) are an integral part of most concurrent systems. An ISR can be considered as a special type of process if certain precautions are taken. In a real environment ISRs usually have a real time response constraint which dictates that they be short and not subject to delays in response. An example of integrating ISRs into a system will be given in Chapter 7.

We must consider two situations. In this section we examine the situation where a producer process passes a buffer for transmission to a consumer ISR. And in Section 2.2.4 we examine the situation where a producer ISR accumulates a buffer which it passes to a consumer process. The situations are essentially the same as that in Section 2.2.1 except for two factors:

1. An ISR cannot be made to wait on a semaphore because of real time response constraints (and possibly deadlock problems). For example, a deadlock occurs if a process executing in a semaphore-protected critical region is interrupted by an ISR which waits on the same semaphore.

2. The output ISR will not be activated unless the physical device has been activated, so a special startup mechanism is required if the I/O device is idle.

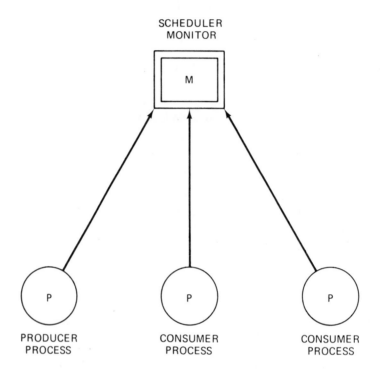

SCHEDULER
MONITOR

M

PRODUCER
PROCESS

CONSUMER
PROCESS

CONSUMER
PROCESS

*Figure 2.2: Access Graph of an Operating System with a Scheduler Monitor*

The first factor does not mean that ISRs cannot access critical regions; it simply suggests that a different mechanism is required to protect processes and ISRs from each other. In a single processor environment where many processes and ISRs wish to access a critical region, the processes protect themselves from each other via semaphores, and processes and ISRs protect themselves from other ISRs by disabling all critical interrupts. In this way the ISRs need not wait on anything before accessing the critical region because the very fact that the interrupt occurred indicates that the critical region is free. In the multiprocessor case, the combination of a Test-and-Set operation and interrupt lockout will provide the necessary mutual exclusion without the danger of deadlock.

In the first situation (i.e., producer process to consumer ISR), the condition "all buffers full" is easily handled as before; the producer process simply waits on a semaphore which is signalled by the consumer ISR when it is done with the buffer, thus freeing space in the buffer queue for another full buffer. There remains the problem of synchronization with respect to the condition "all buffers empty". In this example, this implies the consumer ISR has nothing more to do. In Section 2.2.1 the consumer waited on a semaphore. When the consumer is an ISR, however, it cannot so wait. Instead the ISR will simply become idle since no further interrupts will be received. This problem can be resolved in two ways.

A real time clock ISR could periodically check the status of the associated device and then send the next character if it is idle and another character is in the buffer. This maintains the separation between processes and ISRs, but clearly adds more interrupts to the processor.

The second solution is to create a startup procedure which is called by the producer process whenever the device is idle and a full character buffer is ready. The startup routine, for example, could send the first character to the device thus guaranteeing interrupts for the rest of the buffer. Also, of course the process must be executed by the same microprocessor as the ISR. This solution does not maintain modularity between processes and ISRs, but it can be very efficient.

Figure 2.3 illustrates the mechanism. The queue of full buffers is interrupt-lockout protected. There is no "all buffers empty" semaphore where the ISR waits. Instead a startup procedure is called by the producer processor when it attaches a full buffer while the consumer ISR is idle.

### 2.2.4  Interrupt Service Routine to Process

In the second situation (i.e., producer ISR to consumer process) synchronization with respect to the condition "all buffers empty" can be handled as though the ISR were a process, since the producer ISR only signals the relevant semaphore (as in Section 2.2.1).

There remains the problem of synchronization with respect to the condition "all buffers full". In this case the producer ISR has run out of empty buffers to fill. There are two ways of handling the problem.

The producer ISR can be provided with an overflow ring buffer which allows it to get ahead of the rest of the system to a certain extent. However, the problem is only deferred: what if the ring buffer is overwritten? Then the higher levels of the system must detect errors in the input stream. However, if the ring buffer is large enough the chances of this happening are small.

Alternatively, the producer ISR may throttle further input by disabling interrupts. It then remains for the consumer process or some other active module to reenable the device interrupt at the appropriate time; in effect this amounts to startup of the device, as for the producer ISR to consumer process example. In this case errors may occur due to missed data between the time of disabling the interrupt and restarting the device.

The latter solution is shown in Figure 2.4 which illustrates a complete access graph for the low level software of a keyboard/display console, incorporating both process to ISR and ISR to process communication. It also incorporates ISR to ISR communication as discussed in the next section.

### 2.2.5  ISR to ISR

In some environments it is necessary or at least expedient to have an ISR communicate directly with another ISR.

Figure 2.4 shows an example of ISR-to-ISR communication. The display ISR starts up the keyboard ISR when it displays a prompt for new input. The keyboard ISR starts up the display ISR when it has a character to be echoed and the display ISR is Idle. And the display ISR picks up characters for echoing from an interrupt- lockout-protected echo buffer. Obviously both ISRs must run on the same processor.

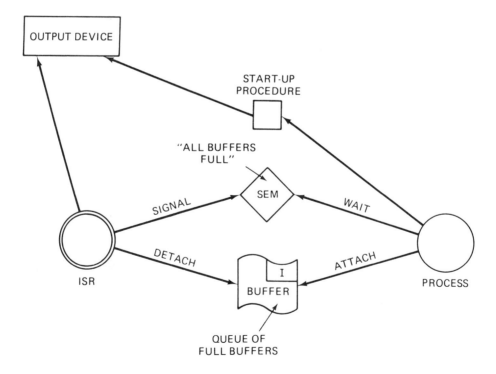

*Figure 2.3: Producer Process and Consumer ISR*

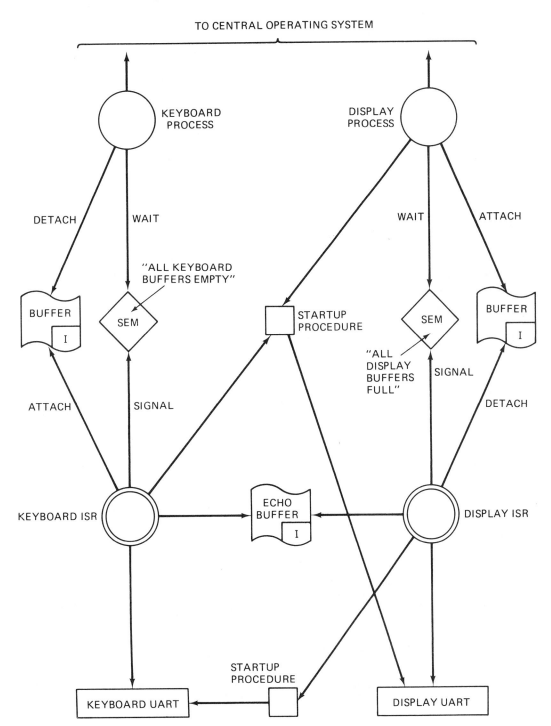

*Figure 2.4: A Console Access Graph*

The need for this intricate low level interaction is caused by our decision, for illustrative purposes, to pass characters to the higher level a line at a time, in emulation of an efficient line-oriented DMA device.

Clearly this approach violates the modularity and hardware transparency principles enunciated earlier. It thus represents a trade-off to achieve fast response. Indeed with typical keyboard display response requirements this compromise is not required and a simpler character-at-a-time interface between the keyboard ISR and the keyboard process is adequate.

### 2.2.6   Some Hardware Transparency Considerations

As we have demonstrated, the care and handling of interrupts is an aspect of any design which tends to destroy attempts at hardware transparency. Solutions are almost always compromises in which the integrity of design principles is violated.

In a multiple microprocessor system it is often difficult to maintain processor access to all peripherals. Thus the processes which interact with ISRs become bound to processors, often early in the design cycle. This loss of transparency creates reverberations which influence other parts of the software design. At this point the effect is merely noted; further discussion of this will occur in Chapter 5 and in Chapter 7.

### 2.2.7   Handlers

In later examples, device handlers will be used to encapsulate the low level mechanisms discussed in Sections 2.2.3, 2.2.4 and 2.2.5 in modules which are seen by the interfacing processes as monitors. These handler modules are in fact degenerate monitors in the same sense that ISRs are degenerate processes. A detailed example of a communications handler monitor is provided in Chapter 7. Here we simply summarize the required features of handlers evident from the examples of Sections 2.2.3, 2.2.4 and 2.2.5:

— Entry procedures which request device services and which are externally similar to monitor entry procedures are required.

— Critical region protection for entry procedures is provided by interrupt lockout or test-and-set operations.

— Interrupt service routines (ISRs) are provided to drive the devices.

— Internal synchronization is via semaphores where possible and via flags and device startup procedures where not (because ISRs cannot wait on semaphores).

## 2.3   An Intelligent Terminal — Example

To illustrate the mechanisms for interprocess synchronization a plausible access graph for the higher levels of an intelligent terminal is shown in Figure 2.5.

The complete system can be separated into three reasonably autonomous operating systems: the central, the communications and the console. They are isolated by monitors which perform the synchronization and scheduling.

The console operating system was discussed in Section 2.2. A communications operating system will be designed and used in Chapter 7 as a vehicle for coalescing the details of concurrent systems. It will not be discussed further here. The central operating system acts as an interface between the operator and the communications operating system. The command interpreter monitor is a scheduler. It receives command data from the keyboard process, interprets it and allocates it to command processes for execution. Each command process then requests and releases buffer space from a buffer manager for data to be sent or received from remote peripherals. Requests are posted with the display request monitor for appropriate operator messages and with the mailbox monitor for external action as appropriate.

All inter-module communication is via buffer pointers which are passed from process to process via monitors. In some cases, as with the mailbox and display request monitors, the monitor is simply a message drop. The command interpreter on the other hand interprets the message buffers, performs actions on them and passes new message buffers on to processes.

The system as shown in Figure 2.5 raises several issues which deserve further exploration. It could be modified to echo characters by passing them through the keyboard process to the command interpreter and thence to the display request. Such a scheme would preserve the modularity of the system at a slight increase in overhead as suggested in Section 2.2.5. In most situations the response would satisfy human requirement for feedback. And the system may have significant interprocess coordination problems, particularly when a command cannot be satisfied because of a system failure of some kind, because of the multiplicity of command processes. This issue is explored further in Chapter 7.

## 2.4   Initialization and Dispatching

## 2.4.0   Introduction

A question naturally arises at this point of how the whole system gets started. Assume for discussion that upon power-up (or reset) the kernel will initialize and start all processes. To start a process means that it will be placed in the ready to run queue as described in Chapter 1. In order to install each process the process control block-(PCB) must be examined to obtain all pertinent data.

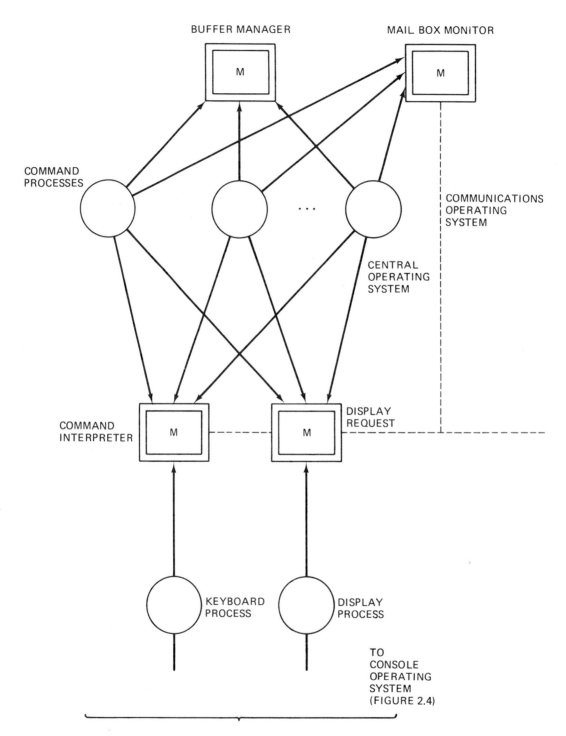

*Figure 2.5: Access Graph of an Intelligent Terminal*

### 2.4.1 Kernel Initialization

If processes are fixed in time then low level initialization code can be made responsible for initializing the kernel and then declaring all processes to the kernel, which then places them in the RTR queue.

This can be accomplished by insuring that on power-up or restart the kernel is locked (by hardware) and the initialization code bypasses this lock. When initialization is complete the kernel is unlocked and regular operation begins.

Note that the initialization code must know about all processes and therefore must be altered if the access graph is changed to include additional processes.

This initialization approach is discussed in Chapter 5.

### 2.4.2 Process Creation

In general kernel primitives could be included in a system which would dynamically create and destroy a set of processes for specific functions. This implies the specification of a PCB and appending the process to the ready to run queue.

## 2.5 Summary

The primary concern of this chapter has been to illustrate the use of semaphores to accomplish interprocess communications. Clearly, the critical regions created by the semaphore can be used to enforce a discipline which should guarantee the desired cooperation. In addition, extensive use of access graphs has been made. This pictorial display of components is a useful technique for grasping quickly the logical interaction of the various system components.

The coordinationg activities within critical regions can become extensive. Indeed they can quickly assume a character of their own. In the next chapter monitors will be introduced and discussed as an example of such a logical component.

## 2.6 References

An early and fundamental reference on processes is:

J.J. Horning and B. Randell, "Process Structuring", ACM Computing Surveys, Vol. 5, No. 1, March 1973, pps 5-30

Chapter **3**

# Monitors

## 3.0 Introduction

The term *monitor* has a variety of meanings. It was originated by early operating systems designers to describe the single monolithic piece of software which handled all concurrency, interprocessor communication and physical resource allocation functions. In our use of the term, a system may have many monitors of varying complexity to handle these functions in a distributed fashion. Monitors in a concurrent system can range from a simple critical region to a complex scheduling and control algorithm.

It is the purpose of this chapter to explore the structure of monitors primarily as reflected in their scheduling mechanisms. In Section 3.1 the general nature of monitors is reviewed and the scheduling problem exposed.

In Section 3.2 a classification scheme based on scheduling is presented. The various monitors are then illustrated with pseudo code in Section 3.3. Limitations are exposed and discussed.

Section 3.4 discusses issues surrounding monitors, including the interface between processes, monitors and the kernel. Interface problems arise in implementations using high level languages and care must be taken to ensure that the critical region begins and ends as planned.

In Section 3.5 a proof of scheduling integrity is presented to provide further insight into the nature of monitors. This section could be skipped on the first reading.

## 3.1 Resource Protection and Scheduling

### 3.1.0 The Nature of Monitors

A monitor is a generalized critical region of code in which a shared resource is manipulated by different processes according to some scheduling rules. This code may be thought of as having two logical parts. The first is the algorithm for manipulating the resource. The code to implement the algorithm is in general indistinguishable from any other sequential code and is written in accordance with the same principles. The second part is the mechanism for scheduling the order in which the associated processes can access the resource.

Establishing a critical region by means of semaphores has been discussed in Chapter 1 and illustrated in Chapter 2. Within this region the strict sequential execution of an algorithm is assured. If the critical region is separated from the process code as a set of procedures, it becomes possible to code, debug and maintain it as a distinct entity; the result is a monitor. The resulting modularity is generally a desirable system attribute.

If the use of a resource requires that processes be scheduled according to some criteria then the overall algorithm must reflect this in some form of testing of the conditions imposed by these criteria. It was proposed in Chapter 2 that entry and exit to the monitor be made transparent to the process code for enhanced modularity. Clearly also, scheduling mechanisms should be transparent for the same reason.

In the next section the scheduling problem is introduced and a simple example used to motivate a full scale discussion of monitor types based on these mechanisms.

### 3.1.1   A Scheduler Monitor Example

Consider once again the producer/consumer example used in Chapter 2. Assume there is only one producer and one consumer. Assume that both the producer and consumer call a scheduler monitor with a single entry procedure REQUEST; a single parameter indicates the nature of the request as follows (the requestor is indicated in brackets):

— get an empty buffer (producer)

— return an empty buffer (consumer)

— get work (consumer)

— put work (producer)

The critical region is implemented using a semaphore called the monitor gate semaphore.

Within the monitor two conditions are monitored:

— empty-buffer-available

— full-buffer-available

```
TYPE MONITOR-PARAMETER=(GET-EMPTY-BUFFER, GET-WORK, RETURN
                          EMPTY-BUFFER, PUT-WORK);
VAR GATE, EMPTY-BUFFER-AVAIL, FULL-BUFFER- AVAIL : SEMAPHORE;
EMPTY, FULL: BOOLEAN;
EMPTYCOUNT, FULL COUNT: INTEGER;
PROCEDURE REQUEST(PARAM : MONITOR-PARAMETER);

    BEGIN WAIT (GATE);

       CASE PARAM

GET-EMPTY- BUFFER:BEGIN IF NOT EMPTY THEN BEGIN EMPTY COUNT: =
    EMPTY COUNT+1; WAIT-AND-SIGNAL (EMPTY- BUFFER-
    AVAIL,GATE);
    EMPTYCOUNT:=EMPTYCOUNT-1 END
    DETACH BUFFER AND MARK EMPTY FALSE IF REQUIRED;
    SIGNAL (GATE)
    END;

GET- WORK: BEGIN IF NOT FULL THEN BEGIN FULL COUNT:=FULLCOUNT+1;
    WAIT-AND-SIGNAL (FULL-BUFFER- AVAIL, GATE);
    FULLCOUNT:=FULLCOUNT-1 END
    DETACH BUFFER AND MARK FULL FALSE IF REQUIRED;
    SIGNAL (GATE)
    END;

RETURN-EMPTY-BUFFER:BEGIN ATTACH BUFFER;
    EMPTY:=TRUE;
    IF EMPTYCOUNT  <>0 THEN SIGNAL (EMPTY-
    BUFFER-AVAIL)
    ELSE SIGNAL (GATE) END;

PUT-WORK: BEGIN ATTACH BUFFER;
    FULL:=TRUE;
    IF FULLCOUNT  <>0 THEN SIGNAL (FULL-
    BUFFER-AVAIL);
    ELSE SIGNAL (GATE) END;

END;
```

*Figure 3.1: A Scheduling Monitor*

These conditions are represented by boolean variables and associated semaphores which are used to force a process to wait if a condition is not valid. It is not sufficient to use semaphores alone to represent the condition because care must be taken to SIGNAL the gate semaphore before being suspended by a WAIT on the condition semaphore (otherwise deadlock would result). Therefore the WAIT cannot be used to test for the condition. And therefore the SIGNAL by itself cannot be relied upon to perform the correct wakeup operation (What if no one is waiting?); separate counters are required.

In the pseudo code of Figure 3.1 the required scheduling is presented. In this code the construct

```
WAIT-AND-SIGNAL (SEM1, SEM2)
```

encapsulates into a single primitive the operations of waiting on SEM1 and signalling SEM2, where SEM1 is the condition semaphore and SEM2 the gate semaphore.

Separation of these operations is both logically and operationally unsatisfactory. It is logically unsatisfactory because the SIGNAL must occur before the WAIT, which is the reverse of the natural way of thinking about the operation. It is more natural to think of blocking the running process before opening the monitor gate. And separation is operationally unsatisfactory because it is inefficient. It forces the execution of two separate kernel primitives and excludes the possibility of combining them into one for efficiency.

The code of Figure 3.1 is unnecessarily complex for so simple an example. Greater clarity and modularity are obtainable by encapsulating the scheduling mechanisms in scheduling procedures with clearly defined external effects. With this end in mind the next section provides a classification of monitors based on their scheduling capabilities.

## 3.2  Classification of Monitors

### 3.2.0  A State Model

In the previous section one type of monitor was illustrated. However several types are possible, with different advantages and disadvantages for each. It is the purpose of this section to provide a classification and comparison of monitor types.

One way of categorizing monitors is by their scheduling capabilities. It is possible, as described in Section 3.1, for a process to begin executing in the monitor and then to find that some condition prohibits completion. In such a situation the process must be blocked in a condition variable to await the condition. A variety of such scheduling mechanisms has been found useful. In the following, four types will be explained.

In order to present a consistent picture of these various monitors, a state diagram will be used. Monitor types are distinguished by different state diagrams. Four different monitor state diagrams can be described using five states:

*WAIT:* Process is waiting on the Monitor Gate
*ACTIVE:* Process is executing monitor code
*BLOCKED:* Process blocks itself waiting for condition to become true
*PENDING:* A high priority waiting state
*ELIGIBLE:* An intermediate waiting state between BLOCKED and ACTIVE

Transitions between states can be described by four primitives. Variations exist in their implementation; however, the four which include definitions by other authors are:

*ENTER:* Request entry to the monitor (i.e., to pass the monitor gate); results in entry to the WAIT or ACTIVE states.
*EXIT:* Relinquish control of monitor to the highest priority waiting process (excluding processes in the BLOCKED state).
*BLOCK:* Temporarily relinquish control of the monitor to the highest priority waiting process; enter the BLOCKED or PENDING states.
*UNBLOCK:* Move a process to the ACTIVE or ELIGIBLE states.

The different monitors have different primitives for blocking and unblocking processes as follows:

For BLOCK: SLEEP, STALL, DELAY, SUSPEND
For UNBLOCK: CONTINUE, WAKEUP, RESTART, PROCEED

The subtle difference between these will be left until their use in various monitors is explained.

Before describing the monitors, it is worth recalling once again that the simplest monitor is a critical region as shown in Figure 3.2 In this case the monitor code is always executed until finished and upon exit a new process enters (or the monitor gate is opened). The state graph shows two states in which a process may be during a monitor call. ENTER implies a wait on the semaphore which controls the critical region. The calling process is either in the WAIT state, suspended on the gate semaphore or it is in the ACTIVE state, implying it is now executing "in the monitor". EXIT implies a signal on the gate semaphore and the dotted line indicates that a waiting process enters the monitor as a result of EXIT by an active process.

### 3.2.1  Monitor of Type Monitor (from Concurrent Pascal)

Consider the state graph shown in Figure 3.3. This monitor was originally used in Concurrent Pascal. In the ACTIVE state, a process can invoke three scheduling procedures:

*DELAY:* This causes a WAIT on a condition variable. It also signals the monitor gate.

*CONTINUE:* This is a special exit procedure which is invoked if the calling process has completed an action which another process may be waiting on (in the blocked state). The blocked process is allowed to return to active. If no processes are blocked, the monitor gate is opened.

*EXIT:* This procedure signals the monitor gate and leaves the monitor. This procedure is the normal exit from a critical region.

This monitor is a simple extension to a critical region in which conditions can be checked before exit. It is simple and the scheduling is easy to understand and implement. It is the type illustrated informally in the example of Section 3.1.1. It has several limitations.

1. The continue must be the last executable statement since an unblocked process will begin to run.

2. Only one process at a time can be continued even though several conditions may have been satisfied.

### 3.2.2   Monitor of Type Manager

An extension of the previous monitor is shown in the state graph of Figure 3.4. In this example, an active process can RESTART any number of processes which immediately leave the monitor. The active process remains in the ACTIVE state until EXIT or SUSPEND.

This mechanism suffers from the constraint that a restarted process must immediately leave the monitor. This may not be appropriate since further monitor code may be required before exiting. For example, there may be some difficulty in returning parameters (such as a buffer address) to a process. This will be explored later.

### 3.2.3   Monitor of Type Mediator (Due to Hoare)

The state graph shown in in Figure 3.5 overcomes the difficulties of the previous two. An additional pending state is added so that a process can schedule others without being forced to leave the monitor permanently. The scheduling procedures are Proceed, Stall and Depart.

*Proceed:* This procedure moves the caller into the PENDING state and unblocks a BLOCKED process which becomes active eventually and leaves the monitor. The Depart primitive restarts a pending process or opens the monitor gate.

*Stall:* This procedure blocks an active process and then looks for PENDING and WAITING processes, in that order. If there are none, the monitor gate is opened.

*Depart:* This procedure is like EXIT except it looks for PENDING processes before looking for WAITING ones.

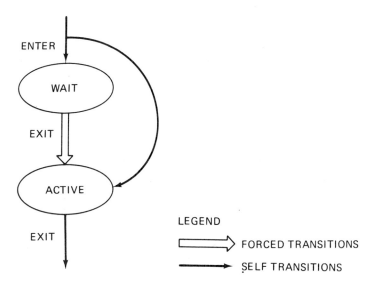

Figure 3.2: A Critical Region

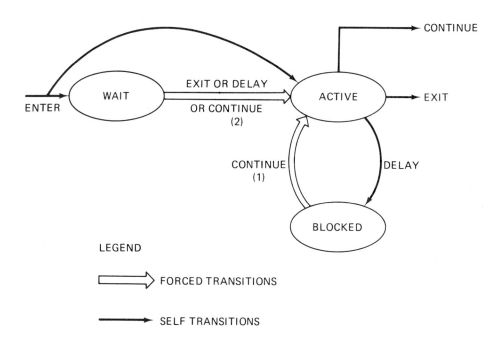

Figure 3.3: Monitor of Type Monitor

While this monitor appears to have the advantage of the first two without the disadvantages, a pathology has crept in; especially for large monitors. The Proceed procedure is invoked at a consistency point (see definition below) by the active process. It then relinquishes control of the monitor. Upon returning it must re-examine the consistency criteria in case changes have occurred.

*Consistency Points:*

> In general, a program has a set of points at which its data is said to be consistent. For a process to release or gain control of a monitor, a consistency point must have been reached and a consistency assertion tested. Proceed and Stall therefore must be invoked only at consistency points. For larger monitors this is rather more difficult than it sounds.

Pragmatically, a monitor must provide the required scheduling and in addition have a minimal number of consistency problems. This implies either the consistency assertions are easy to define and test or that the possibilities for creating inconsistencies

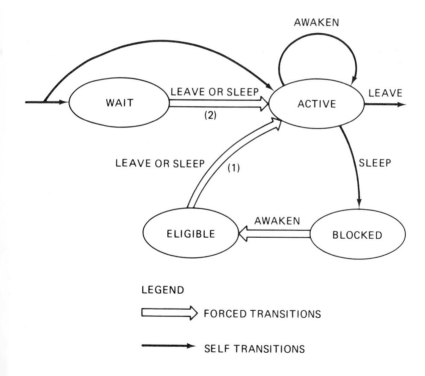

*Figure 3.6: Monitor of Type Gladiator*

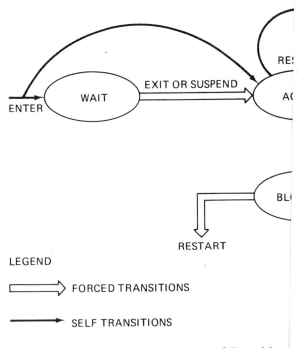

RESTART

LEGEND

⟹ FORCED TRANSITIONS

⟶ SELF TRANSITIONS

*Figure 3.4: Monitor of Type Mana*

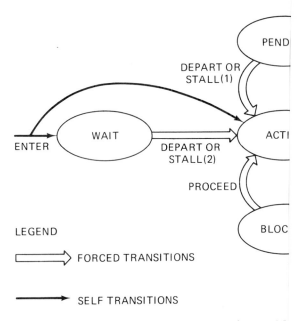

LEGEND

⟹ FORCED TRANSITIONS

⟶ SELF TRANSITIONS

*Figure 3.5: Monitor of Type Me*

over procedure calls are minimized. The next monitor has only one internal scheduling procedure which has a consistency constraint. It would appear to combine the advantages of all the previous types.

### 3.2.4   Monitor of Type Gladiator (due to Cavers and Brown)

Figure 3.6 shows the graph for a monitor which has the logical simplicity of Manager and the flexibility of Mediator (and with only one internal critical procedure).

An ELIGIBLE state is added into which blocked processes move on WAKEUP. The ACTIVE process continues execution. SLEEP or LEAVE cause an eligible process to become ACTIVE (or if none are waiting the monitor gate is opened).

WAKEUP is obviously safe since the active process retains control of the monitor; handover only occurs at SLEEP or LEAVE. Thus, monitor data will not change across AWAKE and data is not required to be consistent at any point other than SLEEP and LEAVE. The only critical internal procedure is SLEEP.

An example of proving the integrity of monitor scheduling for this monitor is given in Section 3.5.

### 3.3   Examples of Monitors

### 3.3.0   Introduction

In this section, pseudo code will be presented to illustrate the implementation of the four monitors discussed in Section 3.2. The Producer/Consumer example will be used.

As before, the Producer and Consumer processes pass pointers to buffer areas. However, the monitor is now required to handle multiple producers and consumers. The producers give the consumers full buffers and return with empty buffers. The consumers get full buffers and return empty ones.

### 3.3.1   Monitor

The scheduling structure of a Monitor of type monitor was shown in Figure 3.3. The data structures for the monitor are shown in Figure 3.7. They consist of two queues of buffer pointers, Full-Work-Queue and Empty-Work-Queue, each with an associated condition variable, Waiting-for-Work and Waiting-for- Return. The monitor scheduling procedures are shown in Figure 3.8. It is assumed that the kernel provides primitives WAIT, SIGNAL and WAIT-AND-SIGNAL for use by these procedures. Recall that WAIT-AND-SIGNAL was defined in Section 3.1.1. Note that the monitor gate semaphore is not passed as a parameter to the monitor scheduling procedures; it is assumed here that monitors have their own copies of these scheduling procedures which know the names of the correct gate semaphores. The code for the monitor entry procedures is shown in Figure 3.9.

The structure is relatively compact and the monitor well suited to this type of scheduling. However, recall that the restrictions are:

1.  An active process can CONTINUE only one blocked process.

2.  An active process must leave the monitor on a CONTINUE.

These restrictions make the scheduling somewhat awkward in the example because dalayed processes cannot leave the monitor as soon as a buffer becomes available but must wait for the continuer to leave the monitor first.

The monitor of type manager using RESTART can start blocked processes without leaving the monitor. This will be illustrated in the next section.

```
TYPE FULL-WORK-QUEUE = RECORD
        COUNT: QUEUE-SIZE;
        QUEUE-AREA: ARRAY [COUNT] OF FULL-WORK-BUFFERS END
TYPE EMPTY-WORK-QUEUE = RECORD
        COUNT: QUEUE-SIZE;
        QUEUE-AREA: ARRAY [COUNT] OF EMPTY-WORK-BUFFERS END
TYPE CONDITION = RECORD
        COUNT: INTEGER;
        SEM: SEMAPHORE END

VAR WAITING-FOR-WORK, WAITING-FOR-RETURN: CONDITION;
        GATE: SEMAPHORE;
        WORK: FULL-WORK-QUEUE;
        RETURN: EMPTY-WORK-QUEUE;
```

*Figure 3.7: Monitor Data Structures for Comsumer Example*

```
PROCEDURE ENTER;
        BEGIN WAIT (GATE) END

PROCEDURE EXIT;
        BEGIN SIGNAL (GATE) END

PROCEDURE DELAY (VAR C:CONDITION);
        BEGIN C.COUNT:= C.COUNT+1;
        WAIT-AND-SIGNAL (C.SEM, GATE);
        C.COUNT:= C.COUNT-1 END

PROCEDURE CONTINUE (VAR C:CONDITION);
        BEGIN IF C.COUNT   <>0 THEN SIGNAL (C.SEM)
          ELSE SIGNAL (GATE) END
```

*Figure 3.8: Monitor Scheduling Procedures*

```
PROCEDURE PUT-WORK (VAR POINTER : ↑ WORK-BUFFER);

   BEGIN ENTER;
     WORK.COUNT := WORK.COUNT + 1;
     WORK.QUEUE-AREA [WORK.COUNT] := POINTER;
     IF RETURN.COUNT = 0 THEN DELAY (WAITING-FOR-RETURN);
     POINTER:=RETURN.QUEUE-AREA [RETURN.COUNT];
     RETURN.COUNT:=RETURN.COUNT-1;
     CONTINUE (WAITING-FOR-WORK)
   END

PROCEDURE FIND-WORK (VAR POINTER: ↑ WORK-BUFFER)
   BEGIN ENTER;
     RETURN.COUNT := RETURN-COUNT + 1;
     RETURN.QUEUE-AREA [RETURN.COUNT] := POINTER;
     IF WORK.COUNT = 0 THEN DELAY (WAITING-FOR-WORK);
     POINTER := WORK.QUEUE-AREA [WORK.COUNT];
     WORK.COUNT := WORK-COUNT-1;
     CONTINUE (WAITING-FOR-WORK);

   END;
```

*Figure 3.9: Monitor Entry Procedures*


### 3.3.2 Manager

A monitor of type manager schedules according to the state graph in Figure 3.4. An active process can restart blocked processes without relinquishing the active state. Now suspended processes can be restarted immediately a buffer is available.

   The data structures are as before. The scheduling procedures are only slightly changed, as shown in Figure 3.10. ENTER and EXIT are unchanged. The code of SUSPEND is the same as that of DELAY except that WAIT-AND-SIGNAL must be followed by an immediate return from the monitor, as indicated by the comment. RESTART differs from CONTINUE in not signalling the gate semaphore; it also performs the counter decrementing which cannot be performed by the restarted process after its return from WAIT-AND-SIGNAL.

   Plausible entry procedures are shown in Figure 3.11. They accomplish the stated purpose but suffer from a fatal flaw. The restarted processes do not leave the monitor immediately as required but stay around long enough to pick up a parameter and then clean up the internal monitor data structures. The problem is that while this is going on, more than one process may be active (the restarted process and the process performing the restart), which is illegal. This type of monitor is obviously too restrictive for applications requiring parameters to be returned to restarted processes.

```
PROCEDURE ENTER;
    BEGIN WAIT (GATE) END

PROCEDURE EXIT;
    BEGIN SIGNAL (GATE) END

PROCEDURE SUSPEND (VAR C:CONDITION);
    BEGIN C.COUNT:=C.COUNT+1;
      WAIT-AND-SIGNAL (C.SEM,GATE) END |AND THEN LEAVE|

PROCEDURE RESTART (VAR C:CONDITION);
    BEGIN IF C.COUNT  <>0 THEN
      BEGIN C.COUNT:=C.COUNT-1;
        SIGNAL(C.SEM) END;

END
```

*Figure 3.10: Manager Scheduling Procedures*

### 3.3.3  Mediator

The scheduling for Mediator was shown in Figure 3.5. To implement this monitor, scheduler procedures shown in Figure 3.12 are assumed. An extra condition variable is required to implement the pending state. As for the gate semaphore, it is here assumed the scheduling procedures for a monitor know the name of the monitor's pending condition variable as well as that of the gate semaphore. The entry procedures are shown in Figure 3.13. They are identical to those of Figure 3.11 except that SUSPEND, RESTART and EXIT are replaced by STALL, PROCEED and DEPART; now the entry procedures are not only plausible but correct because it is not possible for two processes to be active in the monitor at the same time.

However there is still a disadvantage with this type of monitor which makes itself felt particularly in complex monitors. When PROCEED is called in the middle of a monitor entry procedure, as in Figure 3.13, another process takes over the monitor temporarily. Careful programming is required to ensure that assumptions made by the first process about the state of the monitor's data structures are not invalidated by the second process.

```
PROCEDURE PUT-WORK (VAR POINTER: ↑WORK-BUFFER);
    BEGIN ENTER;
        WORK.COUNT := WORK.COUNT + 1;
        WORK.QUEUE-AREA [WORK.COUNT] := POINTER;
        RESTART (WAITING-FOR-WORK);
        IF RETURN.COUNT = 0 THEN SUSPEND (WAITING FOR RETURN);
        {BUT NOW THE NEXT STATEMENTS ARE ILLEGAL}
        POINTER := RETURN.QUEUE-AREA [RETURN.COUNT];
        RETURN.COUNT := RETURN.COUNT - 1;
        EXIT;
    END

PROCEDURE FIND-WORK (VAR POINTER: ↑WORK-BUFFER);
    BEGIN ENTER;
        RETURN.COUNT := RETURN.COUNT + 1;
        RETURN.QUEUE-AREA [RETURN.COUNT] := POINTER;
        RESTART (WAITING-FOR-RETURN);
        IF WORK.COUNT = 0 THEN SUSPEND (WAITING-FOR- WORK);
        {BUT NOW THE NEXT STATEMENTS ARE ILLEGAL}
        POINTER := WORK.QUEUE-AREA [WORK.COUNT];
        WORK.COUNT := WORK.COUNT - 1;
        EXIT;
    END;
```

*Figure 3.11: Plausible but Incorrect Manager Entry Procedures*

```
PROCEDURE ENTER;
    BEGIN WAIT (GATE) END

PROCEDURE DEPART;
    BEGIN
        IF PENDING.COUNT <>0 THEN BEGIN PENDING.COUNT:=
        PENDING.COUNT.1; SIGNAL (PENDING.SEM) END
        ELSE SIGNAL (GATE)
    END

PROCEDURE STALL (VAR C:CONDITION);
    BEGIN C.COUNT:=C.COUNT+1;
        IF PENDING.COUNT <>0 THEN
        BEGIN PENDING.COUNT:=PENDING.COUNT-1;
            WAIT-AND-SIGNAL (C.SEM, PENDING.SEM) END
        ELSE WAIT-AND-SIGNAL (C.SEM, GATE)
    END

PROCEDURE PROCEED (VAR C:CONDITION);
    BEGIN IF C.COUNT <>0 THEN
        BEGIN C.COUNT:=C.COUNT-1;
            PENDING.COUNT:=PENDING.COUNT+1;
            WAIT-AND-SIGNAL (PENDING.SEM, C.SEM)
        END
    END
```

*Figure 3.12: Mediator Scheduling Procedures*

```
PROCEDURE FIND-WORK (VAR POINTER: ! WORK-BUFFER);
    BEGIN ENTER;
          RETURN.COUNT:=RETURN.COUNT+1
          RETURN.QUEUE-AREA [RETURN.COUNT] :=POINTER;
          PROCEED (WAITING-FOR-RETURN);
          IF WORK.COUNT = 0 THEN
          STALL (WAITING-FOR-WORK);
          POINTER := WORK.QUEUE AREA [WORK.COUNT];
          WORK.COUNT = WORK.COUNT-1;
          DEPART
    END;

PROCEDURE PUT-WORK (VAR POINTER: ! WORK-BUFFER);

    BEGIN ENTER;
          WORK.COUNT := WORK.COUNT + 1;
          WORK.QUEUE-AREA [WORK.COUNT] := POINTER;
          PROCEED(WAITING-FOR-WORK);
          IF RETURN.COUNT=0 THEN STALL (WAITING-FOR- RETURN);
          POINTER:=RETURN.QUEUE-AREA [RETURN.COUNT];
          RETURN.COUNT:=RETURN.COUNT-1;
          DEPART

    END
```

*Figure 3.13: Mediator Entry Procedures*

### 3.3.4   Gladiator

The scheduling logic for Gladiator was shown in Figure 3.6. The corresponding scheduling procedures are given in Figure 3.14. As with the Mediator, an extra condition variable, known to the procedures, is required to implement the eligible state. Note the importance of adjusting all affected counts before any signals are performed. For example, in the AWAKEN procedure the eligible count is incremented before signalling the process to be awakened because we know the awakened process will be active in the monitor briefly in procedure SLEEP before waiting on the eligible semaphore. If the count were adjusted by the awakened process in procedure SLEEP, inconsistent results could occur. This point is explored further in Section 3.5.

The key feature of this Monitor type is the relative freedom of the placement of the AWAKEN call. Because no other process is activated by this call until the process performing the AWAKEN relinquishes control of the monitor by calling SLEEP or LEAVE, there is less danger of data structure inconsistency. In the entry procedures FIND-WORK and PUT-WORK of Figure 3.13, AWAKEN, SLEEP and LEAVE may be directly substituted for PROCEED, STALL and DEPART without further modification. However, the reverse is not necessarily true. For example, no harm could result in PUT-WORK from calling AWAKEN (WAITING-FOR-WORK) immediately after calling ENTER before the WORK queue has been updated. However, calling PROCEED at this point would result in an error. In general, the Gladiator type provides greater safety for the programmer.

In Chapter 7 a packet monitor will be presented using the Gladiator type.

```
PROCEDURE ENTER;
     BEGIN WAIT (GATE) END

PROCEDURE LEAVE;
     BEGIN
          IF ELIGIBLE.COUNT <>0 THEN
             BEGIN ELIGIBLE.COUNT:=ELIGIBLE.COUNT-1;
                SIGNAL (ELIGIBLE.SEM) END

                ELSE SIGNAL (GATE)
          END

PROCEDURE SLEEP (VAR C:CONDITION);
     BEGIN
          C.COUNT:=C.COUNT+1;
          IF ELIGIBLE.COUNT <>0 THEN
          BEGIN
          ELIGIBLE.COUNT:=ELIGIBLE.COUNT-1;
          WAIT-AND-SIGNAL (C.SEM, ELIGIBLE.SEM) END
          ELSE WAIT-AND-SIGNAL (C.SEM, GATE);
          WAIT (ELIGIBLE.SEM)

          END

PROCEDURE AWAKEN (VAR C:CONDITION);
     BEGIN
          IF C.COUNT <>0 THEN
          BEGIN C.COUNT:=C.COUNT-1;
          ELIGIBLE.COUNT:=ELIGIBLE.COUNT+1;
             SIGNAL (C.SEM)
          END
     END
```

*Figure 3.14: Gladiator Scheduling Procedures*


## 3.4 Issues

### 3.4.0 Introduction

A complete discussion of issues surrounding the design, implementation and test of monitors will be reserved until Chapters 7, 8 and 9. Chapter 7 presents a complete system design example. Chapter 8 shows how monitors and complete concurrent systems may be implemented using available high level languages. Chapter 9 discusses a number of issues relating to both design and testing. We comment here on a few issues which the reader may already be worried about. The brief discussion here paves the way for further discussions in subsequent chapters. The issues are as follows:

— Monitor/Process Interface design

— Monitor/Kernel Interface design

— Monitor Integrity

— Monitor Testing

— Monitor Initialization

## 3.4.1   Monitor/Process Interfaces

In Section 3.3 examples of monitor scheduling were presented to illustrate the implementation of the types introduced in Section 3.2. In addition to choosing and implementing an appropriate scheduling mechanism for a particular application it is also necessary to design an interface. This is because the point at which a critical region is established must be clearly identified to avoid errors in execution and to guarantee the integrity of parameters. Ideally the process code should call procedures and not be involved in details of scheduling or entry. This is most easily accomplished if the language used for implementation supports reentrancy.

The problem is that the first high level language statement of our example is the ENTER call which locks the gate, but low level code may be executed before this call to pass parameters. In PASCAL there is no problem because procedures are automatically reentrant. In PL/M, which is a version of PL/1 supported by INTEL Corp. for its range of microprocessors, there may be a problem if reentrant code is not used. One way of getting around this problem is as follows:

In the simplest case the process could lock the monitor gate:

> Call Enter;
> Call Find-Work (pointer);

This approach suffers the disadvantage of a loss of modularity, and some clarity. It places the lock and unlock mechanisms in different systems modules, which is undesirable. A reentrant entry procedure is illustrated in Figure 3.15 which avoids this.

If reentrant PL/M code is deemed too inefficient then a reentrant entry procedure "shell" may be used to pass parameters to the monitor and lock the gate. This reentrant shell then calls on an internal monitor procedure which is not reentrant and which does the work. This approach is illustrated in Figures 3.16 and 3.17.

In these PL/M examples a different approach to condition variables is used which will be explained in the next section.

```
MONITOR: DO;
DECLARE ALL PRIVATE DATA
----
/*ENTRY PROCEDURE*/
FIND$WORK:  PROCEDURE (POINTER) REENTRANT PUBLIC
DECLARE POINTER ADDRESS;
DECLARE WORK BASED POINTER ADDRESS;
/*ASSUME A KERNEL CALL TO A PRIMITIVE ENTER */
CALL ENTER;
---
IF WORK IS WAITING
      THEN DO;
          WORK=WORK PARAMETER  BLOCK POINTER;
          CALL EXIT;
          RETURN;
          END
CALL DELAY (.WAITING$FOR$WORK);   /*KERNEL CALL*/
END FIND$WORK;
----
END MONITOR
```

*Figure 3.15: Reentrant Entry Procedure*

### 3.4.2  Monitor/Kernel Interface

In the examples of Section 3.3 the monitor/kernel interface was assumed to consist of simple calls to a WAIT, SIGNAL or WAIT-AND-SIGNAL procedure. In practice, however, the kernel is likely to be a permanently located piece of assembly language code, perhaps located in ROM, with a single entry point. It is inconvenient at best to link the kernel to all monitor modules.

An alternative approach, using PL/M, is shown in Figure 3.18 for a monitor of type monitor. Here the kernel entry point from a monitor is at absolute address "CHLL". As will be described later in Chapter 5, this single entry point provides for kernel functions shared by all kernel procedures (mutual exclusion, context switching, etc.). The numeric codes in the kernel calls designate primitives and semaphore parameters, as will be explained in Chapter 5.

A different approach to condition variables is illustrated here which will be explained further in Chapter 5. Instead of using an augmented semaphore as the condition variable, the kernel primitives BLOCK-AND-SIGNAL and UNBLOCK simply mark processes as blocked or ready-to-run in the process descriptions. And the scheduling procedures simply store and read the process index in a location inside the monitor whose address is provided in the DELAY and CONTINUE calls. To facilitate this mechanism the kernel stores the index of the currently running process on each processor in a special private memory location called NAME which may be accessed by the monitor scheduling procedures running on that processor. The monitor itself must then maintain the condition variable queues. This approach has the advantage of provided freedom of choice of queue disciplines for condition variables, but the disadvantage of requiring one additional kernel primitive UNBLOCK. The primitive BLOCK-AND-SIGNAL replaces WAIT-AND-SIGNAL, which is no longer needed.

If the kernel provides only WAIT and SIGNAL but not WAIT- AND-SIGNAL (or BLOCK-AND-SIGNAL), then the monitor scheduling procedures are changed only slightly. For example in Figure 3.14 only the SLEEP procedure is changed, as shown by Figure 3.19. Note the extra kernel call required.

```
/*Producer Process*/
----
CALL PUT$WORK (PTRI, PTR2)
----
/*MONITOR ENTRY*/
PUT$WORK:   PROCEDURE(REQUEST$PTR,PUT$ANSWER$AT) REENTRANT
              PUBLIC;
DECLARE LOCAL VARIABLES
CALL   ENTER;
CALL INS$PUT$WORK(REQUEST$PTR,PUT$ANSWER$AT);
CALL DELAY (.Q (WAITING$LOC)):/*WAIT FOR ANSWER*/
END PUT$WORK;
```

*Figure 3.16: Reentrant Shell in PL/M*

```
/*Internal Monitor Procedure*/

INS$PUT$WORK:   PROCEDURE (REQUEST$PTR,PUT$ANS$AT);
----

/*IF the correct process is waiting for work, put the request directly
in the expected slot and restart the process. ELSE put the work in the
correct Queue. The consumer will pick it up when it reenters the
monitor*/

IF WAITING$FOR$WORK = NIL;
THEN DO
      Q(ENTRY) = REQUEST$PTR;
      Q(FLAG)  = 1;
END;
ELSE DO;
---
WORK = REQUEST$PTR;/*PASS WORK DESC.*/
CALL RESTART (.WAITING$FOR$WORK);/* WAKE UP CONSUMER*/
END;
END INS$PUT$WORK;
```

*Figure 3.17: Internal Procedure Called by Reentrant Shell in PL/M*

```
            DECLARE KERNEL LITERALLY 'CHLL';
                    NIL LITERALLY '-2';
            DECLARE   NAME BYTE EXTERNAL;
            DECLARE   WAIT LITERALLY '0312H';
            DECLARE   SIGNAL LITERALLY '0412H';
            DECLARE   BLKSIG LITERALLY '0512H';
            DECLARE   UNBLK LITERALLY '0600H';

        KERNEL: PROCEDURE (BC$REGISTERS, DE$REGISTERS) EXTERNAL;
            DECLARE (BC$REGISTERS,DE$REGISTERS) ADDRESS;
        END KERNEL;

ENTER: PROCEDURE;
        CALL KERNEL (WAIT,0); /*WAIT(GATE)*/
END ENTER
EXIT: PROCEDURE
        CALL KERNEL (SIGNAL,0); /*SIGNAL(GATE)*/
END EXIT;

DELAY: PROCEDURE (ADDRESS$OF$QUEUE);
        DECLARE ADDRESS$OF$QUEUE   ADDRESS;
        DECLARE QUEUE BASED ADDRESS$OF$QUEUE BYTE;
        QUEUE=NAME;
        CALL KERNEL (BLKSIG,0); /*BLOCK AND SIGNAL(GATE)*/
END DELAY
CONTINUE: PROCEDURE (ADDRESS$OF$QUEUE)
        DECLARE ADDRESS$OF$QUEUE ADDRESS;
        DECLARE QUEUE BASED ADDRESS$OF$QUEUE BYTE;
        DECLARE TEMPORARY BYTE;
        TEMPORARY = QUEUE;
        QUEUE = NIL;
            IF TEMPORARY  <>NIL THEN;
            CALL KERNEL (UNBLK+TEMPORARY,0);/ *UNBLOCK(PROCESS)*/
        END CONTINUE
```

*Figure 3.18: Monitor Calls to Kernel in PL/M*

```
PROCEDURE SLEEP (VAR C:CONDITION);
  BEGIN

      C.COUNT:=C.COUNT+1;
      IF ELIGIBLE.COUNT   <>0 THEN
      BEGIN
      ELIGIBLE.COUNT:=ELIGIBLE.COUNT-1;
      SIGNAL(ELIGIBLE.SEM) END
      ELSE SIGNAL (GATE);
      WAIT (C.SEM);
      WAIT(ELIGIBLE.SEM)

  END
```

*Figure 3.19: SLEEP Using Simple Semaphore Primitives Only*

### 3.4.3   Monitor Integrity and Testing

An important consequence of coalescing scheduling operations within a monitor is the increase in importance of the monitor integrity. Fortunately, the monitor structure is ideal for self testing. The monitor can be considered as a finite state machine and the monitor code as implementing that machine. It is an easy matter to insert extra processes which wait on internal conditions and keep a journal of events. This journal can be typed immediately, or on a timed basis or recalled in the case of failures.

During initial debugging a sequence of test points in the monitor can be defined. As such points are reached the test process can be passed a validation parameter to display. The monitor can wait until the display is finished, then continue. In the event of malfunction a complete record is available.

During initial operation, a ring buffer can be substituted for the display to maintain a sequence of snapshots which can be examined in the event of malfunction to discover more subtle pathologies. The design of such a test bed will be presented in Chapter 9.

Further insight into monitor integrity is provided in Section 3.5.

### 3.4.4   Monitor Initialization

Monitors must include among their entry procedures special initialization procedures which may be called by startup and restart software. These procedures may initialize all monitor variables or only those associated with restart of some sub- function of the monitor. Further insight is provided in Chapters 7 and 8.

### 3.5   Monitor Integrity

The design of the scheduling functions of a complex monitor can be approached by using an FSM as has been demonstrated. The detailed operations are the subject of specific verification; however, the scheduling mechanisms can often be proved to be correct. The major concerns are of course that only one process executes monitor code at a time and that the waiting counts are always consistent. As an example of such proof the monitor of type Gladiator is examined based on the code in Figure 3.14. A four-step argument due to Cavers and Brown establishes the required results:

1) *If* only one process at a time executes monitor code *then* the waiting counts for eligible and for all condition variables are consistent on entry to and exit from each of the four scheduling procedures.

*Proof*: The counts are consistent because the single process executing the relevant scheduling procedure code adjusts them to correct values before performing any signal.

2) *If* the waiting counts are consistent, *then* no excess signals will accumulate against the eligible or condition semaphores.

*Proof*: None of these semaphores is signalled unless the waiting count is non-zero. The count field of the semaphore itself, therefore, can be positive only during the transient intervals following a signal and preceding the wait executed by a process whose intention of doing so was declared by incrementing the waiting count.

*Corollaries*

1. A process executing sleep is always suspended on the condition semaphore until a wakeup is later executed.

2. An awakened process is always suspended on the eligible semaphore until the monitor is released.

3. It follows that wakeup does not increase the number of processes executing monitor code.

3) *If* the waiting counts are consistent *then* at most one process can execute monitor code at a time.

*Proof*: Note first that only sleep and exit could be suspected of increasing the number of processes (from corollary 2 of assertion 2). But in both of these, the process releasing the monitor either hands it over to exactly one other process or increments the count field at the gate semaphore. Hence the sum at the gate count field and the number of processes executing is constant. Since the gate count field is initialized to one, the assertion must hold.

4) The waiting counts are consistent and at most one process at a time executes monitor code.

*Proof*: Assertion three shows that if they are consistent, then at most one process at a time executes monitor code, and that this restriction on the number of processes guarantees continued consistency (which follows from assertion 1). Since the waiting counts are initialized to zero they are and will remain consistent. The assertion is thus proven.

Such a proof is not easy to compose; however, attention must be given to such verifications or interesting bugs can appear which could be difficult to detect, find and eliminate. A further assertion is possible from the previous ones, i.e.:

5) *If* there are processes waiting to run, *then* no new processes can enter the monitor.

*Proof*: This follows from the implementation of sleep and wakeup.

As a result, this data could be made inconsistent in the period between a wakeup and the running of the awakened process only by the process performing the wakeup, or by another process in the eligible queue which is ahead of the process under consideration.

## 3.6  Summary

Monitors are key components in the access graph which describes a concurrent system. They perform the dual role of synchronizing the access of concurrent processes to a resource and the manipulation of that resource. A process, while operating "in the monitor" is in a critical region which can be designed, debugged and guaranteed correct. The modularity of the resulting system is excellent and the complexities of concurrency remain transparent at the process level with respect to those functions performed by the monitor. Of course in a many-process, many-monitor system, the process programmer must be concerned about higher level concurrency problems.

The state graph representation of monitors is a convenient and useful way of displaying the scheduling properties. For complex monitors it is easy to see the major features and to direct the coding and debugging. In addition this state approach can be utilized as a test bed for debugging and maintenance by introducing test conditions which call test processes.

Parameter passing to the monitor depends on the facilities of the language used. Reentrancy is a convenient and foolproof way of providing for multiple calls while preserving the integrity of the passed parameters and the transparency of the system.

## 3.7  References

Concurrent Pascal is described by:

PER BRINCH HANSEN, "The Programming Language Concurrent Pascal", IEEE Trans. Software Engineering, Vol. SE-1, No. 2, June 75

The fundamental paper on monitors is:

C.A.R. HOARE, "Monitors: An Operating System Structuring Concept", CACM, Vol. 17, No. 10, Oct. 74

# Chapter 4

# Hardware — the Inner Circle

## 4.0  Introduction

In general the hardware for microprocessor based systems is expanding rapidly, and it is not the purpose here to discuss components in detail. Good references are widely available and will be quoted where appropriate.

At the core of the "onionskin" systems model proposed in Chapter 1 is the hardware. It is our purpose in this chapter to examine the concepts and issues surrounding the choice of multiple processor architectures. These issues are complex, and interrelated to such an extent that optimal configurations are often difficult to perceive. Two extremes exist in a particular design environment. On one hand the whole question of hardware is open, and on the other a fixed or partially fixed configuration forms the target. In the latter case tuning is required to map the software onto the hardware. In the former, the results of an integrated system design procedure must yield an optimal hardware structure. Such procedures are not widely known; however, it is clear that good choices do not proceed out of ignorance. And an extensive knowledge of hardware is required to create viable systems.

In addition to the knowledge of hardware normally associated with the microprocessor field, the most outstanding feature of multiple processor systems is their interconnection topology. Generally, in conventional systems a processor is connected to a set of peripheral components in a master/slave relationship. This configuration is widely familiar, and includes interface chips with intelligence rivalling the processor. We assume here a familiarity with the components of microprocessor systems. It is with the attributes and characteristics of a wide variety of interconnected structures that we are concerned.

In Section 4.1 a general discussion of the "why" of multiple processors will be presented. This leads to a requirement for a classification scheme. Widely accepted classification schemes are not available and a discussion of the issues and variations concludes the section.

In Section 4.2 the interconnection of components is considered in a general way. The topological structures are examined in terms of certain attributes which affect

**65**

performance. The requirement for interconnection leads in a natural way to a discussion of bus structures and in particular to arbiters and protocols. These are examined in Section 4.3.

In Section 4.4 a common bus configuration as an example of a tightly coupled system is explored in more detail. Such systems are commercially available and their characteristics and performance are of particular interest.

Finally, the requirements are focused on the processor and a "wish-list" of desirable characteristics is formulated based on hardware considerations.

## 4.1  Distributed Computer Systems

### 4.1.0  Introduction

There is no universally accepted taxonomy to define, identify or classify distributed computer systems. However, there have been noticeable trends even in early uniprocessors to decentralize certain functions, e.g. input/output channels. Historically there has been a constant demand to lower the cost performance ratio of computers. The mechanisms have depended on a subtle balance between the cost of hardware and our ability to create, implement and support innovative architectures. In this section we shall expand on the "why" of multiple processor systems and present a simple classification scheme.

### 4.1.1  Why Distributed Systems

Except for special application areas, the enhancement of computer performance has generally been achieved by refinements of the basic concepts proposed by Von Neuman. Thus the IBM 370 is not a startling architectural departure from his original computer. Advances in the performance of semi-conductor components have had more influence on performance than any other single factor. Over the last few decades the major thrust was to increasing the raw computing power of large machines. The price/performance for most applications favoured such machines. Until recently, the hardware/software balance was such that a collection of small computers, equivalent in processing power to a large computer, cost more than the large computer; given that they could be interconnected and controlled. This balance is changing; even so, the motivation for using multiple processors does not come from throughput considerations only, but from the many other advantages of distributing both processing power and intelligence. These advantages include:

*Geographical or Functional Modularity of Application*

Many systems are geographically distributed in their functions without the need for tight centralized synchronization. Pre-processing of data, local autonomy over response tactics and lower communications costs often lead to a lower overall systems cost with enhanced performance.

*Reliability*

Redundancy is easier to create in distributed systems. Graceful degradation as a component fails is often an important system specification.

*Response*

Human interaction and other real time situations often demand a time response which is relatively simple but critical. Local intelligence usually can meet such response requirements more easily than a centralized system.

*Cost Modularity*

The interconnection topology of distributed systems can be designed to allow incremental growth and even static or dynamically reconfigurable systems.

*Resource Sharing*

Hardware resources as well as programs, data bases and even computational power can be shared (rather than duplicated) in a distributed system.

*Software Implementation*

Software is often easier to design, implement, debug and maintain if it is functionally partitioned and executed on dedicated processors.

*Friendly Interfaces for Non Professional Users*

Many systems are now being used by operators with little or no computer training; indeed with no interest in computers. They demand a friendly, intelligent interface with the system. This function can be most economically provided by local intelligence dedicated to being "friendly".

The bottom line is a cost-performance figure which, in terms of present trends and for certain applications, is in favour of distributed function architectures.

Some recent (1978) results indicate clearly the cost and performance trends in multiple processor systems.

The multiple processor (PDP-11) system at Carnegie-Mellon has been analysed by Fuller. He compares four bench mark programs against a DEC PDP-10 (a conventional uniprocessor). He reports a price performance ratio of 3.59 in instructions per second per dollar.

An off-track billing system was reported by Rosenblatt. Here an MPS of mini-computers yielded four times the throughput of two 360/50 systems at a rental cost of $250,000/month less.

Cornell reports on a multiple processor system designed to operate a telescope which searches the sky for satellite debris. A throughput rate of 9 million instructions/sec is required. The system calls for ten minicomputers in a common bus configuration. It was estimated that the total cost will be 10% of a suitable uniprocessor.

McGill and Steinhaff compared an IBM 370/168 to a multiple processor system consisting of a mini and ten micros. For an aerial combat simulation program they report a 500/1 improvement in cost-effectiveness.

Finally, Wirshing projects to the 1985 time frame. He considered problems with a high degree of parallelism (e.g. weather forecasting). For uni-processors designed with present architectural concepts he predicts a 40 to 1 cost performance advantage for multiple processor systems.

Of course not all problems will show this kind of performance advantage. It would be folly to preclude uniprocessor solutions as a first consideration. For the advantages of multiple processor systems do not all come easily. Careful designs are required to realize a given cost performance specification: e.g.

*Distributed Systems can be inefficient:*

Interprocessor synchronization and communication reduces the effective power of the processors. This overhead can limit the total throughput as more processors are added. The selection of processors, the allocation of functions to them, the interconnections topology, the organization of the data structure, the system control strategy, all are important design issues in this regard.

*Development of Software:*

The development of an operating system can increase the costs of software above that of a conventional uniprocessor. Extensive software for testing processors in isolation and in concert is often required. Debugging concurrent systems can be an exceedingly complex (and frustrating) task. Microcomputer software support is not yet as extensive as for standard processors.

*Man Power:*

Design teams need a high degree of interrelated expertise in microprocessor hardware, interfacing, real time programming, communications and concurrent operating systems. Such people are not widely available.

The specification and description of the desired system in an appropriate way remains difficult. The techniques for partitioning the functions and the hardware are only heuristically understood. Without these problems this book would be unnecessary. And it is towards a design methodology for a large class of such problems that we are progressing.

Before proceeding, a classification of hardware architectures seems appropriate. In the next section a classical scheme is presented of which the multiple processors in which we are interested form one class. The overall classification scheme is simplistic and appealing and has withstood the test of time, despite some limitations.

## 4.1.2  Classifications of Computer Structures

Early computers were classified as serial or parallel, depending on the design of the arithmetic logic unit. In a broader sense, the concept of parallelism may be extended to cover any set of operations carried out in parallel. Thus it may occur at any logical or physical level of the system. As a result, early attempts at classifications tended to mire in conflicting perspectives or definitions. Classification schemes should obviously be inclusive, differentiate between structures, be easy to apply, and be useful. Most proposals fail some or all of these criteria.

A scheme proposed by Flynn has found acceptance in general and with simple extensions it forms a reasonable way to view architectures. He based his scheme on the concepts of *Instruction Stream* and *Data Stream*. These concepts are not rigorously defined; however, a computer executes a sequence of instructions on a sequence of data, regardless of how each arrive at the point of action, Multiplicities in these streams lead to four possibilities:

Single Instruction Single Data — SISD
Single Instruction Multiple Data — SIMD
Multiple Instruction Single Data — MISD
Multiple Instruction Multiple Data — MIMD

Since Flynn's proposal (1966), modifications and extensions to his scheme have appeared. Figure 4.1 shows a general structure with extensions by other authors.

It is not our purpose here to pursue the general area of classifications, although the subject is important and most interesting. We note that the MIMD class tends to describe the architectures in which we are interested. Within the MIMD classification it is possible to differentiate between three sub-classes based on the degree of physical coupling between processors.

### 1) *Loose Coupling*

Generally geographically dispersed with low band-width ($\leq$50 Kbits/sec) interconnections. The ARPA network is such a system. It offers *resource sharing* with the individual processors being independent non-homogeneous units. Loose coupling characterizes *computer networks*.

### 2) *Moderate Coupling*

Here there is an increased level of inter-computer activity with higher speed serial or parallel buses. Typically the processors cooperate to execute an algorithm. They are not general purpose systems but are designed for a particular class of problems. Moderate Coupling could be used to characterize multiple processor systems (MPS).

### 3) *Tight Coupling*

Tight Coupling implies high band width interconnections and hence physical closeness of processors. They generally use shared memory for data transfers and program storage. Such configurations are often called *multiprocessor systems*.

The latter two categories are often difficult to distinguish. A considerable grey area exists which defies precise classification. We are generally interested in Multiple

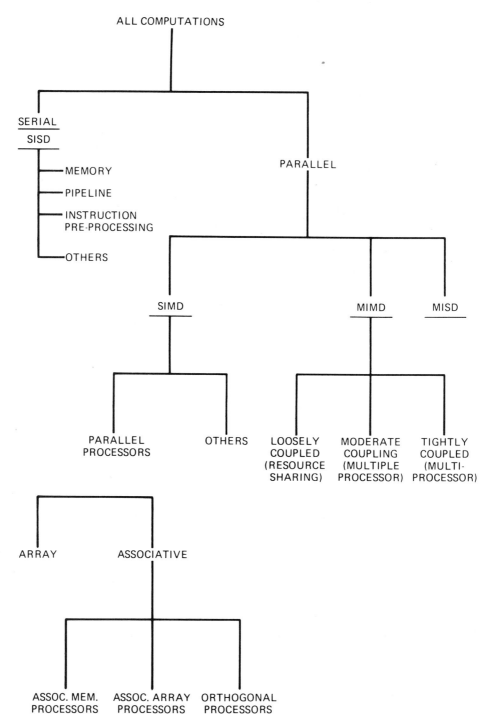

*Figure 4.1: Flynn's Computer Classification with Extensions*

Processor Systems, as defined above, although the techniques to be described extend well into the Multiprocessor area. We are interested as well in the design of systems to form nodes of computer networks. However, the design of the networks themselves is outside the scope of this text.

Multiple processor systems (MPS) exhibit a diversity of architectures. The processors can be homogeneous or otherwise; physically close or dispersed; and interconnected in a variety of topologies. The system tasks can be partitioned and allocated to a fixed processor or could migrate between processors. Control can be central or distributed. This wide diversity in logical and physical implementation demands a design algorithm which systematically examines each alternative and resolves each design issue.

Several global attributes tend to differentiate multiple processor systems and deserve discussion at this time:

*Interconnection Topology:* A wide variety of physical interconnections are possible between processors. The resulting topology influences the modularity of the system as well as creating bottlenecks which may limit performance. Interconnection schemes have been discussed in a clear and explicit fashion by Anderson and Jensen and Ramamoorthy. A common bus is widely used for connecting multiple microprocessors. Such systems have good modularity (i.e. easy to add a processor) but the bus is an obvious bottleneck. A more detailed discussion of interconnection topologies will be introduced in Section 4.3.

*Logical Structure:* Processors in an MPS are either logically equal or hierarchically structured. An hierarchical or vertical organization implies a master/ slave type organization. Such systems are generally easier to build but are less flexible. Logically equal or horizontal systems are more difficult to build and control but are more flexible, reliable and capable of dynamic load sharing.

*Function Allocation:* The functional division of the system is an important step in the overall design process. The subsequent allocation of functions to processors can be done in two ways. Functions can be dedicated and fixed to a processor or allowed to migrate. The dedicated approach simplifies the control software but decreases flexibility and makes dynamic load sharing impossible. The alternative is more complex to control but has the advantages the former lacks.

Despite the appearance of some structure to these classifications it will be observed as we proceed that considerable blurring of the boundaries often occur and systems are described which have the characteristics of several classifications. The classifications thus form a starting rather than a resting point.

## 4.2  Interconnection Topologies

### 4.2.0  Introduction

The interconnection of processors, memories and peripherals distinguishes the various implementation of multiple processor systems. Unfortunately there is no universal scheme for classifying the various topologies which have been or might be used. It remains useful nevertheless to have certain agreements on a taxonomy for discussion purposes. And it is even more important for the designer to know the operational (or functional) implications of the attributes exhibited by a particular configuration.

We will discuss in Section 4.2.1 a set of attributes applicable to any configuration. These attributes are not quantifiable in general and are used as mechanisms for qualitative comparison of alternate interconnection proposals. In comparing specific choices it becomes necessary to rate in some fashion each attribute and perhaps to quantify its effect by analysis or simulation. An example of such an analysis to find a performance limitation (bottleneck) will be given in Section 4.3.2. In Section 4.2.2 a classification scheme which includes a wide variety of existing topologies is discussed. This scheme illustrates in a top-down fashion the implications of various decisions as the components are interconnected. It is valuable from this point of view alone. The common bus is used for more detailed comment; other configurations are examined in Chapter 9.

### 4.2.1   Attributes of Connected Structures

The following attributes can be used to examine different proposals for connected structures. No relative importance is implied by the order of presentation.

1.  Performance Bottleneck

    A bottleneck is a term which describes a feature or component which limits a performance measure. Bus bandwidth for example can limit data throughput. Bottlenecks are thus limitations which can critically affect system performance. The location or nature of the bottleneck can force alternate connections or the duplication of components or other modifications of existing or proposed systems.

2.  Modularity

    Modularity is an attribute which refers to the ease with which incremental changes can be made to the structure or capability of a system. Two measures are often used: Cost and Place modularity.
    Cost modularity refers to the dollar cost of adding a component relative to its cost. If the cost of an addition is the cost of the component only, the system exhibits the ultimate in cost modularity.
    Place modularity refers to limitations on enhancing performance by adding a component because of its location. Performance increases may be difficult to obtain at certain places in a system and easy in others. A system could have good cost modularity but poor place modularity (and the opposite of course).

3.  Fault Tolerance and Reconfigurability

    We refer here not to reliability but to the affects of a fault on the system and its ability to continue functioning in a reconfigured structure.
    Each component in a system exerts some influence in the event it fails. A fault

which can cause system failure is termed a *single point failure*. In the event of single point failures a system may be reconfigurable such that some aspects of system performance are maintained. Or it may be desirable to duplicate these components in order to maintain full system capability in the event of a fault. Other faults may degrade system performance but not cause a complete functional failure. Data flow for example may be rerouted around a failed processor or a function may be suspended (e.g., an Input/Out). The cost of assuring the reliability of system performance measures depends in the first instance on its configuration and second on the single point failure components and finally on the remaining components.

4. Interconnection Complexity

This attribute refers to inherent logical complexity of routing data from one processor to another. It is difficult to quantify. To some extent it is a software problem and refers to the totality of decisions made to establish communications and to the data necessary for these decisions.

Not all of these attributes are important in all situations; however, any interconnection proposal can be examined using this list as a reference. Alternative proposals can be quantitatively compared by assigning suitable values to each attribute. In the future we will refer to this list when discussing interconnection schemes.

## 4.2.2 A Classification Scheme

Attempts to classify the topologies which result from interconnecting components have been made at roughly three levels of activity: logic design, computer architecture, and computer networks. Such classifications attempt to create a semblance of structure by relating design decisions to performance characteristics.

Of the several proposals which characterize interconnection topologies, the following is the most useful for our purposes. More important than any proposed taxonomy is the recognition that decisions form a hierarchy. Thus strategic decisions made early in the process have far reaching implications both on system performance and on implementation.

The simplest perspective of a multiple processor system is a collection of processors which intercommunicate to perform a function. The processors are connected into topological structures by means of paths and switches.

A *path* is the medium by which messages are transferred (e.g., buses, memories, etc.)

A *switch* is on devices for establishing the routing of a message (e.g., physical devices or computers).

Based on this model an attractive top-down approach can be organized based on a four level hierarchy as shown in Figure 4.2. The root of the tree is the decision to connect two or more processors. The first two levels are concerned with policy and the last two with implementation issues.

Strategic Policy

Level 1 — Connection

The first decision is either to directly or indirectly connect source and destination. An indirect connection implies the intervention of one or more switching entities. Switches combine to select the message path. The selection process is based on routing strategies which are implemented jointly or singularly by the switches, possibly based on control information carried in the message.

Level 2 — Routing Controller

Indirect routing strategies can be classified as either central or distributed. A centralized controller establishes a route for each message. Decentralized control implies a routing algorithm executed at each switch.

Implementation Policy

Level 3 — Path Structure

At the third level the paths are designated either as shared or private (dedicated). A shared path implies more than two users. It implies further that contention for the path becomes an implementation issue which must be solved by some form of arbitration. This will be discussed in Section 4.3.1.

Level 4 — System Architecture

At this level the system interconnection becomes identifiable.

As an example the global bus structure will be examined in some detail. In the global bus (Figure 4.3d) all processing elements communicate by means of a common bus. The bus is allocated to a processor pair by an arbiter for the duration of a message (or possibly smaller transfers). Such a structure has the following features:

*Bottleneck*

The bus bandwidth is the obvious bottleneck.

*Modularity*

Both cost and place modularity are good. Processors can be added at any place on the bus with little cost. From the perspective of communications bandwidth, however, the modularity is low. Increasing bus bandwidth usually requires substantial redesign or possibly replication.

*Fault Tolerance and Reconfigurability*

Both are good with respect to processor failures but obviously catastrophic with respect to bus failures. Processor failures can be isolated from the bus by good design; however, bus failures (which are not uncommon) require a duplicate bus if the system is to continue operating.

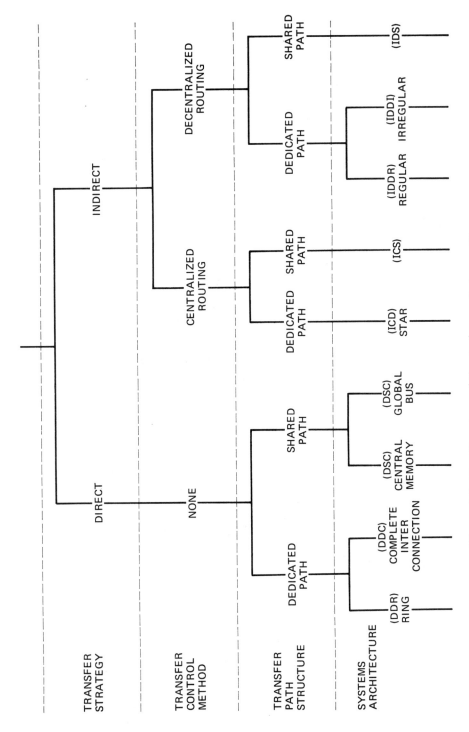

*Figure 4.2: The Structure of the Anderson & Jenson Taxonomy*
Copyright 1975, Association for Computing Machinery, Inc., reprinted by permission.

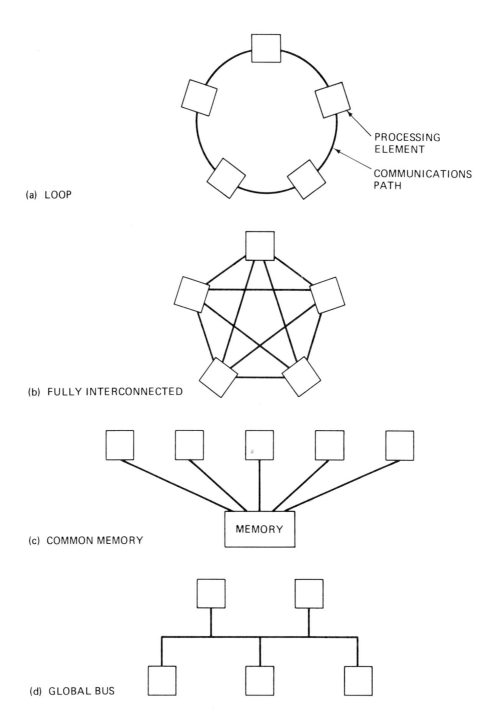

*Figure 4.3: Interconnection Configurations*

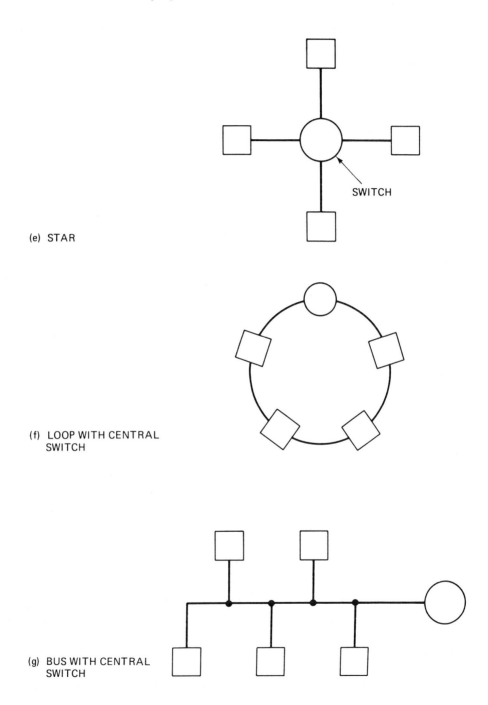

(e) STAR

(f) LOOP WITH CENTRAL
    SWITCH

(g) BUS WITH CENTRAL
    SWITCH

*Figure 4.3: Interconnection Configurations (Cont'd)*

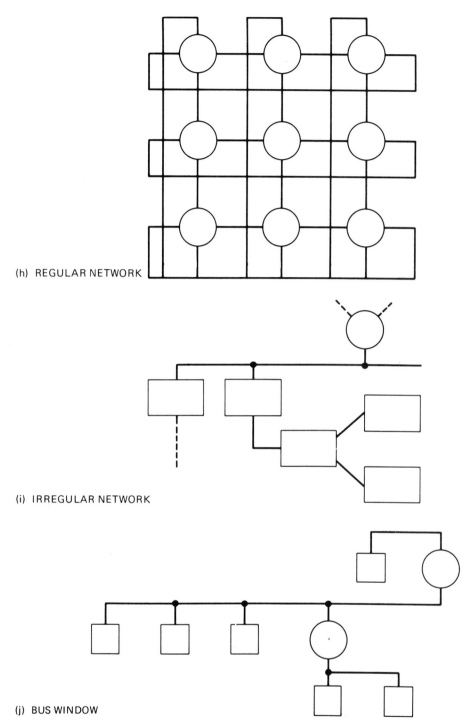

(h) REGULAR NETWORK

(i) IRREGULAR NETWORK

(j) BUS WINDOW

*Figure 4.3: Interconnection Configurations (Cont'd)*

*Connection Complexity*

Low; although an arbiter must be included to resolve simultaneous request for bus usage.

A concise picture of the overall taxonomy is shown in Figure 4.4. It is a broad scheme which encompasses both computer systems architecture and computer newtworks.

Proposals for the classification of multiple processor systems will undoubtedly continue to appear periodically. It is unlikely that a universal scheme will ever be found acceptable. Any such scheme should include all possible topologies in an explicit way and accommodate new innovations as they occur. In addition the scheme should be useful to systems designers and architects. This section has considered general principles only. In the next section ways of implementing interprocessor communication between the components of an interconnected system will be examined in detail.

Transfer strategy:

Direct
    Path: Dedicated
        Topology: Distributed loop control
        Topology: Complete Interconnection
    Path: Shared
        Topology: Common memory
        Topology: Common Distributed Control Bus

Indirect
    Routing: Centralized
        Path: Dedicated
            Topology: Star
            Topology: Loop
        Path: Shared
            Topology: Common Centrally Controlled Bus
    Routing: Decentralized
        Path: Dedicated
            Topology: Regular Network
            Topology: Irregular Network
        Path: Shared
            Topology: Bus Window

*Figure 4.4: Taxonomy of Anderson and Jensen*

## 4.3   Bus Structures and Inter-Processor Communications

### 4.3.0   Introduction

In Section 4.2 nodes were assumed interconnected by paths which provided the total mechanisms required for the transfer of control and data. Such paths could be implemented in diverse ways. In this section we are concerned with the logical structures of such paths and the means whereby information is moved from source to destination. Primarily the discussion concerns wired connections between nodes; however, it is not limited conceptually to such paths. We shall refer to paths as buses in the following.

There are three major issues involved in the design of buses:

1.   Dedicated or Shared

2.   Communications Technique

3.   Data Transfer

A shared path involves two further sub-issues. Simultaneous requests for usage must be arbitrated and nodes must be connected to the bus.

*Arbitration*   is a procedure by means of which nodes request and obtain exclusive use of the bus.

*Connection*   is the mechanism by which the nodes are given physical access to the bus after being designated as the user.

Communication refers to the overall means by which two or more nodes transfer information.

The word *protocol* is used to describe the whole information transfer process, including, in general, request for and granting of access to the bus.

Data transfer refers to the volume of data moved at each access to the bus. This parameter influences the other two. Indeed all three are interrelated and we will first describe each in isolation and then discuss their mutual impact. Much of the structure of this section is due to Thurber et al. Thurber's paper also contains an extensive annotated bibliography.

In the following three sections we shall deal first with bus arbiters, then with protocols and finally with parallel and serial buses.

### 4.3.1   Bus Arbiters

Six alternative arbitration schemes are shown in Figure 4.5. The logical distinction here is between centralized and decentralized control of the arbitration process. The mechanisms in either case for receiving requests and for granting access are called daisy chain, polling, and independent requests. The six alternatives are not exhaustive but they are the most common.

Before discussing the six alternatives, two comparative performance and implementation measures will be proposed. The first is cost and the second arbitration time.

The cost of an arbiter can be measured in two ways, namely by communication costs and by logic costs. Communication costs refers to the number of arbitration signal lines, drivers and receivers

Centralized
>   Daisy Chain
>   Polling
>   Independent Requests

Decentralized
>   Daisy Chain
>   Polling
>   Independent Requests

*Figure 4.5: A Classification of Bus Arbiters*

The logic cost is a weighted measure of the number of critical logic elements such as registers, comparators, etc. It is a measure of the logical simplicity of the arbiter as reflected in the cost of components required for implementation.

Unfortunately these two costs are difficult to quantify exactly. The relative weighting of the logic and communications cost to obtain an overall cost is also difficult to agree upon. Thus cost tends to be somewhat qualitative but in specific cases of comparison it is still useful.

The second measure is total arbitration time, namely the elapsed time from a request to a grant. This time depends on how many potential requests exist. It is often convenient therefore to consider several time measures, the simplest being the time from grant to request if only one request is posted. If simultaneous requests are considered, either the average or worst case times are useful. The choice between centralized and decentralized arbiters will be discussed after a brief explanation of each.

Centralized arbitration schemes are shown in Figure 4.6. They operate as follows:

*Centralized Daisy Chain*

Requests are asynchronously raised on a common request line. The central control propagates a bus grant signal if Busy is false. The first unit which has signalled a bus request that receives the Grant signal stops its propagation, raises the busy line and assumes bus control. On completion, it lowers the busy line and a new Grant signal is generated if further requests are outstanding.

The unit closest to the controller always has the highest priority. Furthermore, if a unit fails to lower the busy signal or fails to propagate the grant signal the system fails.

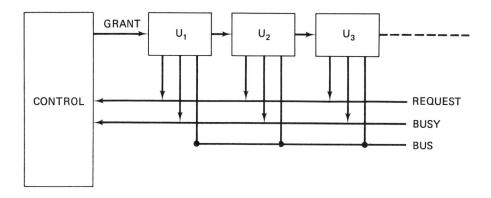

(a) DAISY CHAIN

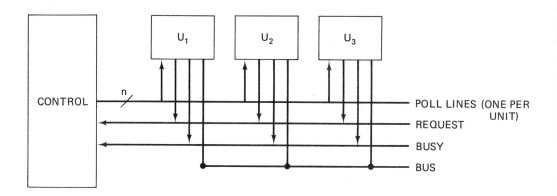

(b) POLLING

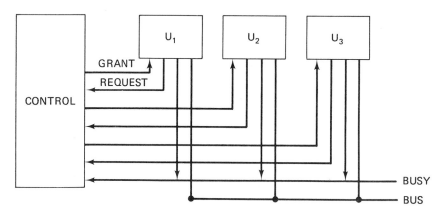

(c) INDEPENDENT REQUEST/GRANT

*Figure 4.6: Centralized Bus Allocation*

## Centralized Polling

In this scheme the grant line is replaced by a set of wires of sufficient width to address all units. On a request the controller sequences through the unit addresses. When a unit requesting access recognizes its address, it raises busy. Polling ceases and the unit has access until it lowers busy. This technique could incorporate dynamic changes in priority allocation.

## Independent Request

This scheme requires 2 lines for each unit, as shown in Figure 4.7. The controller can now execute any programmed allocation algorithm. Clearly it is the most flexible, and dynamic changes in priority allocation can be programmed. A failed unit can be timed-out and not cause systems failure. A single point failure does exist if a unit does not or cannot be forced to drop the busy line.

Decentralized systems imply that control is distributed over each unit. The three varieties are similar to the centralized approach as shown in Figure 4.7.

## Decentralized Daisy Chain

Decentralization is accomplished by creating a circular connection for the Grant line as shown in Figure 4.7. A clock pulse is daisy chained around the loop and a request by a unit blocks this pulse. Thus when a unit raises a request (on an internal flip-flop for example) and it receives a grant pulse, it takes control of the bus. On completion it regenerates the grant pulse which circulates until blocked by another unit. Such a scheme is logically simple and requires only two lines. However, the failure of a unit fails the whole system. Such systems are used on physically compact systems where the signal lines are on a common back plane.

## Decentralized Polling

In a decentralized polling scheme, shown in Figure 4.7, each unit is responsible for transmitting a polling code. Thus a unit requiring access to the bus must wait until it recognizes its code. It then raises busy and uses the bus. On completion it lowers busy and transmits the next polling code.

Each unit is a single point failure since failure to transmit the next polling code causes system failure. Each unit is logically simple and priority can be achieved by altering the poll sequence.

## Independent Requests

In this approach, shown in Figure 4.7, each unit has a bus request line which is assigned a priority. When busy is lowered all units examine all bus requests and the unit recognizing its own priority as highest raises busy and assumes control of the bus. This approach is not subject to single point failures; however, the complexity of each unit is high.

Each of the six configurations are viable depending on the factors which control the selection. They can be compared in several ways.

*Daisy Chains*

This is the simplest structure of all. Only a few control lines are required and these are independent of the number of devices. In the centralized case the controller is logically simple. The simple logical structure and low implementation costs are however affected by several disadvantages:

a) Priority is fixed by physical location in the case of central control and is round robin in the decentralized version.

b) Low Place Modularity unless the units are close together; otherwise long cables may be necessary to insert a new unit into an appropriate priority slot.

c) Failure is catastrophic if any unit fails to function.

d) Assignment is slow due to the ripple delay and is obviously dependent on the number of units in the chain.

Despite these disadvantages Daisy Chains are widely used in physically compact systems because of their simplicity and low cost. Multiple processors on a single backplane for example provide an ideal application. The polling clock rate can be considerably higher than the average microprocessor clock and thus the ripple delay becomes transparent.

*Polling*

Either approach to a polling arbitor requires a considerable increase in control hardware and bus lines with a resulting increase in logical complexity. In the centralized case a flexible priority allocation is possible since polling sequences could be altered dynamically. In both cases a modular system results. In addition a reduced failure susceptibility is evident. A failed unit which does not respond to the poll is ignored. This is an outstanding feature.

On the negative side, the number of units is limited by the number of polling lines. And the polling rate must be slow enough to allow response from an active unit.

*Independent Requests*

This approach compares logically to the interrupt structure of computers. It is potentially the fastest and most flexible in allocation. Failed devices can be ignored which yields a high immunity to unit failures.

It requires a large number of control lines and in either case the controller is complex.

Its speed and flexible allocation potential make it attractive for multiple processor applications.

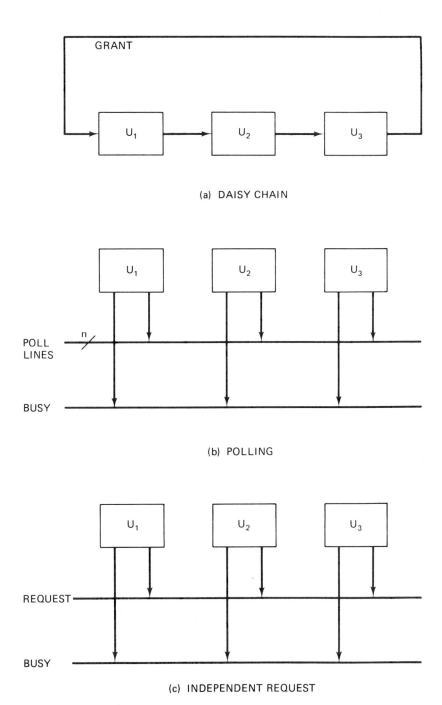

Figure 4.7: Decentralized Bus Allocation

### 4.3.2  Bus Protocols

The word protocol has wide ranging meaning, depending on the logical level of the message transfer procedure being discussed. A multi-layered view of a packet switching protocol will be presented in Chapter 7. At this time we are concerned with the electrical signals on the bus used to establish and maintain data transfers. These signals are often called data link control (DLC) protocols and are at the lowest logical level of a protocol hierarchy. At this level the protocol signals are executed by hardware as part of the bus interface. Specifications for the DLC protocol usually include voltage and current as well as rise times and other pertinent data to define pulses and levels. In general, a programmer writing a high level language transfer process is unaware of the DLC protocols. He has a higher level protocol appropriate to his tasks and requirements. Here we are concerned with a description of the logical aspects of the exchanged signals and how they affect data integrity.

A classification scheme proposed by Thurber is shown in Figure 4.8. The major categories are synchronous and asynchronous. Synchronous communications imply that sender and receiver have an agreement in advance as to timing or some way of arriving at such an agreement when needed. Each unit shares a common clock or has separate clocks which are synchronized (to within a specified tolerance). Asynchronous implies that no such common time information is available.

Measures of performance of any scheme include those proposed for arbiters as well as speed (bus band width) and error susceptibility. The total data transferred includes the bits required by the protocol plus the message.

```
SYNCHRONOUS
ASYNCHRONOUS
     ONE-WAY COMMAND
          SOURCE CONTROLLED
          DESTINATION CONTROLLED
     REQUEST/ACKNOWLEDGE
          NON-INTERLOCKED
          HALF-INTERLOCKED
          FULLY-INTERLOCKED
```

*Figure 4.8: Communication Mechanisms*

Thus the actual transfer of n bits of data requires some m bits of protocol signals. The effective bus band width in message bits per second is lower than the bit transfer rate of the bus. Clearly the protocols are an overhead (albeit necessary) which determines the effective bus band width. There is usually a trade-off between costly protocols and band width. By cost we mean (as with arbiters) the number of signal lines, drivers, etc., required to implement the protocol as well as the logic cost of the components required for implementation.

Electrical protocols are designed to achieve high speed transmission of data consistent with some criteria for minimizing errors. Errors can occur because of noise on the bus or because of potential errors in the protocol itself. There are two possibilities:

a) protocol errors

    i)  missed data

    ii) duplicate data

b) bus induced errors

*Missed data* occurs when the transmitter forces data onto the bus when the receiver is not yet ready to accept it. *Duplicate data* errors occur when a receiver copies the data twice (or more) on the assumption that new data has arrived. *Bus induced errors* (called glitches) result from noise or other spikes which are mistaken as control signals or which distort data signals.

The protocols can be examined using timing diagrams which show the time relationship of the various signals or as state graphs which display the logical intent of the signals; both are useful. The interfaces at the source and sink can be considered as finite state machines which implement the desired protocol. The electrical signals on the bus are used to provide synchronization for the state transitions of the two machines. The extent of the synchronization determines the error potential of the protocols.

In the state graph representation which follows, the transitions are "event triggered". A single connecting line between states indicates that timing is controlled by a private event (perhaps a clock) while a double line indicates a transition triggered by an external event.

Asynchronous protocols fall into two categories — open- ended or one way, and a request/acknowledge exchange.

*One-way protocol*

One way protocols are either source or destination controlled as shown in Figure 4.9 and 4.10. Timing between source and sink is implicit and no acknowledgement of data reception is given. The major problem is insuring an appropriate delay so that data is valid before it is captured. In a fixed installation this timing can be adjusted and a high speed system results. Its simplicity is attractive; however on a shared bus, devices of different speed would demand an adjustment to accommodate the lowest response time.

*Request/Acknowledge*

A request/acknowledge protocol demands an acknowledgement (ACK) from the receiver. And in some cases a negative acknowledgement (NACK) to indicate an error. Three variations are possible depending on the degree of coupling between each end of the bus.

a)  non-interlocked

The timing diagram is shown in Figure 4.10 while the source and sink FSMs are shown in Figure 4.12. The two machines are synchronized only by the ACCEPT signal. Thus either from the FSMs or the timing examples the protocol could cause lost or duplicate data. Duplicate data could occur if the

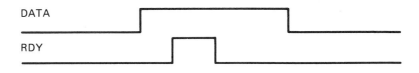

(a)  SOURCE CONTROLLED

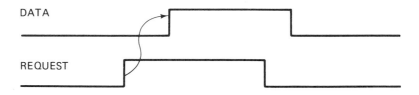

(b)  DESTINATION CONTROLLED

*Figure 4.9: Asynchronous One-Way: Timing*

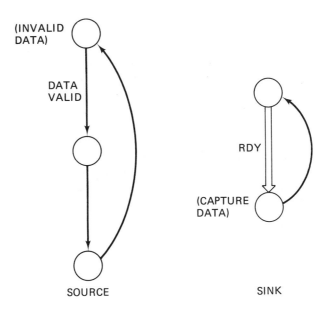

(a) SOURCE CONTROLLED

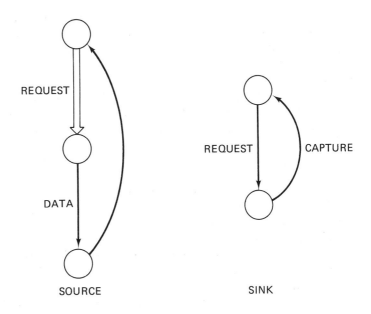

(b) DESTINATION CONTROLLED

*Figure 4.10: FSM for One Way Protocol*

RDY level is held up too long by the source and misinterpreted as a second data transfer command. Data could be missed if the ACCEPT pulse is held up too long and the second RDY command overlooked.

b) half-interlocked

Electrical timing signals are shown in Figure 4.11 while the source and sink FSMs are shown in Figure 4.13. In this case duplicate data errors are avoided since the ACCEPT signal is used to drive both RDY and DATA down. However, lost data is still possible if the ACCEPT masks the arrival of the next READY.

c) fully interlocked

The source and sink FSM are shown in Figure 4.14. Electrical timing is shown in Figure 4.11. In this case missing and duplicate data errors are eliminated since the two machines are now driven by interlocked control signals as shown.

A missing or distorted signal can cause bus hang-up in the all cases. If this possibility is foreseen and depending on failure requirements, a time-out can be used to recover and institute recovery action.

By forcing responses in terms of levels rather than leading or lagging edges, a large measure of noise immunity is also gained. However, it is noted that even without noise the protocol could be inappropriate to the system performance requirements. The more complex protocols automatically adjust to devices of differing speeds which is often the situation on a shared bus.

Synchronous communications implies the agreement by source and destination on a global time standard. Three major issues must be considered; the mechanics of synchronization, data transmission and verification. Of course bus utilization and effective band width must also be traded against reliability.

*Synchronization*

Timing on a bus may be generated either centrally or locally. A centralized system contains a clock which is broadcast to all units on the bus. Obviously this is subject to errors due to skew as distances increase. Typically a framing signal can be propagated which is decoded by each unit and used to drive local counters which then identify the correct time. This framing sequence may be distributed by a separate bus or mixed with data. A separate line would still create the possibility of skew (between it and the data bus).

If the framing information is mixed with the data, it must be separable either by amplitude, phase or coding characteristics.

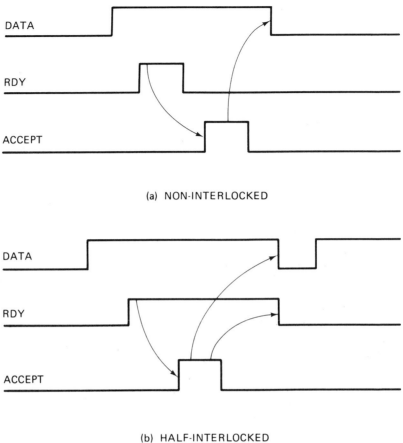

(a) NON-INTERLOCKED

(b) HALF-INTERLOCKED

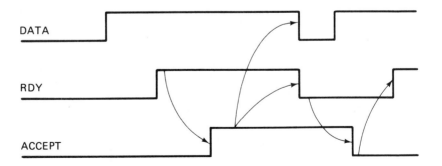

(c) FULLY INTERLOCKED

*Figure 4.11: Asynchronous Request/Acknowledge*

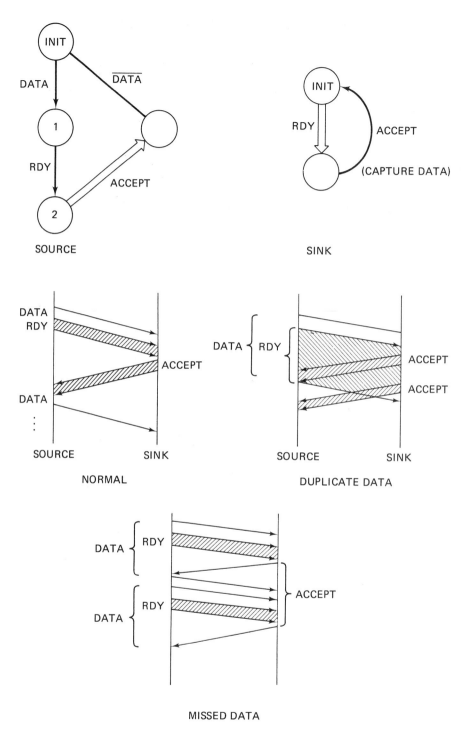

*Figure 4.12: Non-Interlocked FSM and Error Possibilities*

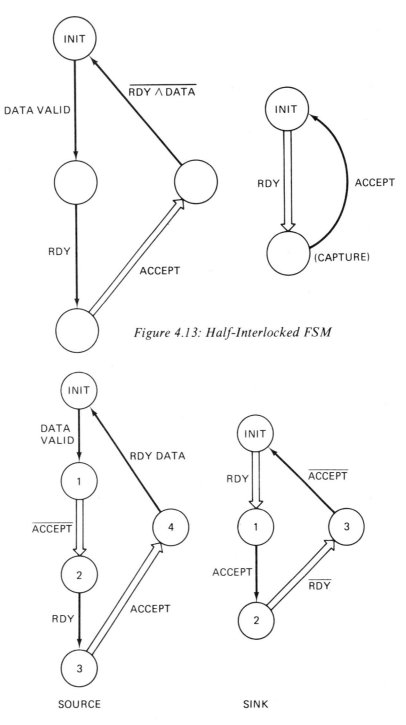

*Figure 4.13: Half-Interlocked FSM*

SOURCE

SINK

*Figure 4.14: Fully Interlocked*

Amplitude coding demands more complex line drivers and receivers than simple binary data. Phase coding usually requires a long sequence which often exceed the time slot allocated to a device; this seriously reduces bus band width. Coded data implies a special sequence which must be detected. Somewhat more complex devices are required to generate the sequence and to insure normal data does not generate an equivalent sequence.

### Data Transmission

Having reached an agreement on time, each unit can be allocated a time internal (slot) during which it utilizes the bus. These slots are allocated on a dedicated or non dedicated basis (and sometimes a mixture).

### Dedicated Time Slots

Allocation on a dedicated basis is used most often when units are relatively homnogeneous in their requirements. Obviously dedicated slots remain permanently allocated no matter how frequently (or otherwise) they are utilized. If capacity or speed discrepancies do exist, a device could be given more than one slot (called super commutation), or by sub-multiplexing, more than one device could be assigned to a single slot.

Clearly if different speed devices must be supported, the choice of the operating speed becomes critical. Buffers must be used to match the slower devices to a high speed bus. And high speed devices must obviously be allocated sufficient time to accommodate their transfer rates.

### Non-Dedicated Slots

Such slots are assigned to devices as they are requested. This improves utilization but clearly requires an allocation algorithm. as well as some means for exchanging requests and grants. A fast bus assignment mechanism would allow single word or block transfers.

### Verification

Verification of correct data by the receiver is usually required to ensure the reliability of the transfer. This requires an acknowledgement to be transmitted back to the sender in the appropriate time slot. In the simplest case each slot could be wide enough for a reply for every word. If the bus contains slow devices, they will limit the band width for such a scheme.

Verification by default is often used which implies a request for retransmission in a designated time slot only if an error occurs.

## 4.3.3   Parallel Buses

A parallel bus implies a simultaneous transfer of a number of bits of data using one of the control mechanisms. This requirement is so widespread that a large number of microprocessor bus peripherals is available. These chips couple directly to the processor bus and on the bus side execute the required protocols for direct transfers in and out of the processor. On the other side they can be programmed to execute a variety of FSMs, depending on the requirements of the system.

The Intel 8255 is such a chip which can be dynamically programmed (others include the motorola PIA). Further discussion of these chips is beyond the scope of this text.

### 4.3.4   Serial Buses

A serial bus usually implies a single wire interconnection between source and sink. The single wire is used for both control and data transfers. The protocols and transmission techniques for serial buses are outside our scope of reference.

### 4.3.5   Data Transfer Philosophies

The volume of data transferred during each connection of source and destination tends to influence decisions on the other factors. Data volume can be assigned as a:

- Single word

- Fixed length block

- Variable length block

For convenience in the following discussions the term efficiency will be used to quantify the data transfer relative to protocol bits as follows:

let $n_c$ be the number of bits to code a character
(including any parity bits)

$n_p$ be the number of protocol bits required to
transmit a message of N characters

then

$$\text{efficiency} = Nn_c/(Nn_c + n_p)$$
$$= n_c/(n_c + n_p/N)$$

In some instances a similar definition using time is more appropriate. Thus efficiency can be measured as the ratio of time required to transmit a character string to the total time required (including the protocol). In either case we note that longer blocks of characters tend to improve the transfer efficiency.

*Single character*

> Transfer protocols must operate at the character level which implies high speed, low bit protocols for efficiency.

*Fixed length blocks*

> Block transfers amortize the protocol overhead over more characters and thus achieve higher efficiency.

*Variable length blocks*

> Most flexible, however, block counts must be included and hence the controller must be more complex.

The communication environment may require one of these choices. Single character transfers are inefficient with efficiencies ranging from 20 — 75%, depending on the control bus structure. Fixed blocks may not be suitable for a number of different devices on a bus; however, efficient transfers are possible. Variable length blocks offer the most flexibility with a more complex controller and usually lower efficiency when compared to the transfer of an equivalent fixed block.

Many schemes are combinations of these; for example, a single word or a fixed sized block.

## 4.4    Tightly Coupled Systems (Common Memory)

## 4.4.0    Introduction

A tightly coupled system will imply throughout this discussion one which is physically compact. And a common memory structure will be assumed as an example. Such systems are widely used and several such configurations are currently available. Commercial processor boards with associated arbiters and backplanes provide complete facilities for implementing such systems. These architectures are characterized by:

1) Low functional complexity

2) Good cost and place modularity

3) Bus bandwidth as the bottleneck

4) Single mode failure potential (the bus or memory)

5) Physical compactness (usually small systems)

6) Relatively low cost

Despite the physical closeness of the components, there is a wide discrepancy in speeds and hence such systems have both synchronous and asynchronous communications capabilities. We shall look first at some common configurations, then discuss arbiters and resource protection.

### 4.4.1  Classifications

For our purposes the four configurations shown in Figure 4.15 will be discussed. There is no widely agreed upon classification scheme and the proposal here covers systems which are available. It is assumed in every case that any component capable of becoming a bus master gains control of the bus as the result of a request/grant sequence with an arbiter. Once in control of the bus, message exchange with bus slaves is in accordance with agreed on protocols.

a)  Global Resources Only

In this configuration all peripherals and memory are included on the common bus. All processors execute code from this memory. The system is physically and functionally homogeneous. The bus is an obvious bottleneck which quickly limits the total systems throughput (in instructions/sec). However the system is modular and homogeneous. With the right operating system, processes can run on any processor and hence processor failures are not catastrophic. Graceful degradation can be built-in. On the other hand single-point failures can occur in both the bus and the memory.

b)  Private and Global Memory

In this configuration the bus traffic is reduced by placing some code in a memory private to each processor. The system in general becomes non-homogeneous unless the code in each private memory is identical. A non-homogeneous system implies that all processes can not run on all processors. Thus a processor failure can impair or halt the system. Increased throughput has been obtained at the expense of an additional failure mode and a loss of reconfiguration capability (and graceful degradation).

c)  Distributed Resource System

In this category the peripherals are assigned permanently to a processor. This is often done in practice but of course all of the disadvantages of the previous category now apply with respect to input/output. On the positive side it is often easier to debug and to trouble-shoot a distributed system than a homogeneous one. This may account for their wide use.

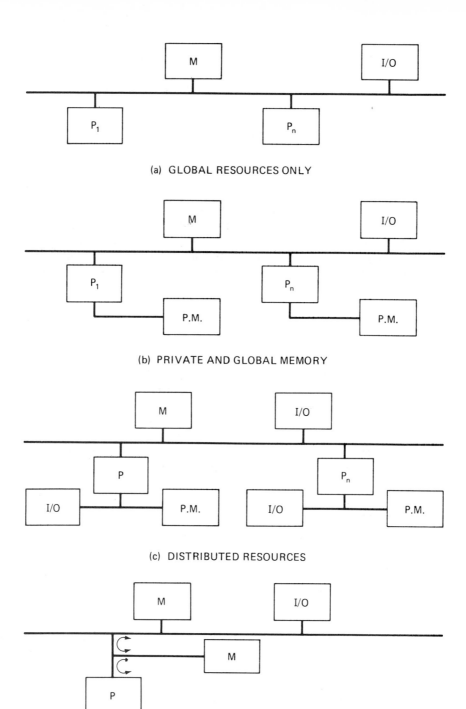

(a) GLOBAL RESOURCES ONLY

(b) PRIVATE AND GLOBAL MEMORY

(c) DISTRIBUTED RESOURCES

(d) DUAL PORT MEMORIES

*Figure 4.15: Common Bus Configurations*

### d)  Dual Port Memories

Dual port memory configurations allow private memories to be accessed from the bus. Thus a memory map can be created in which each processor can view other processors' memories as extensions of their own. This is most useful in distributed systems when a block of data is acquired from a peripheral on one processor which is to be processed by another. Without a dual port memory the data would be transferred to public memory for processing or depending on bus traffic to private memory.

With respect to failure modes, it is most often observed that bus failures predominate over processor failures. Thus the distribution of resources to processors is not as serious as might be envisaged. Indeed, during design, the various processes often fall naturally into partitions which are allocated to a processor.

### 4.4.2  Performance

The common bus configurations discussed in the previous section suffer eventually throughput limitations imposed by the system bus. The configuration in which all memory is common will suffer first the effects of bus congestion. The others will be characterized by the allocation of functions to the various memories. It is of course a design objective to partition the system in such a way that the effects of potential bottlenecks are avoided.

In this section a performance analysis of a homogeneous common memory system is presented which is related to costs. The cost analysis illustrates the trade-offs for this system and is applicable with modification to the other configurations. The relationship of costs to throughput points to some interesting requirements for future microprocessor instruction set design.

Consider N processors (with no private memory) connected to a common bus. Define

System Throughput ($T_s$) — total instructions per second executed by the system

Two factors affect the apparent congestion created by the bus. First is the arbitration delay. In most designs this delay is negligible and will be neglected here. The second is the actual utilization of the bus during normal instruction execution. Define

Bus Utilization ($U_b$) — ratio of bus cycles required by a processor to those available.

Examination of the fetch and execution cycles of typical microprocessors reveals a bus utilization for only a fraction of the total time. Given that no waiting is required on memory, ratios of 0.2 to 0.5 are common.

If no contention exists then the system throughput for N processors should equal N x $T_p$ where $T_p$ is the processor throughput. If $U_b$ of one-third is assumed, then for 1, 2 or 3 processors no contention will exist. This follows since after an initial transient each processor will synchronize into every third slot. Thus $T_s$ will rise linearly with N. However, for $N \geq 3$, a bottleneck has been reached and no further increase in $T_s$ is possible. Thus a curve as shown in Figure 4.16 is typical.

If no interference were encountered the throughput is $NT_p$. This will be decreased as a function of N to a new value $T_m$. Thus

$$NT_p/T_m = \text{throughput with no interference/throughout with interference}$$
$$= (1/U_b)/1/U_b + N\text{-}1$$
$$= 1/(1 + U_b(N\text{-}1))$$

where $1/U_b$ is the number of bus cycles used to execute an instruction and (N-1) is the number of cycles of delay. Curves for $T_m$ are shown in Figure 4.17. Reasonable increases are still possible even with the rather pessimistic assumptions on wait intervals.

It is worth observing here that if different arbitration is implemented that similar curves can be computed; but that some processors on the low end of the priority allocation scheme may be effectively excluded from executing any instructions. Thus

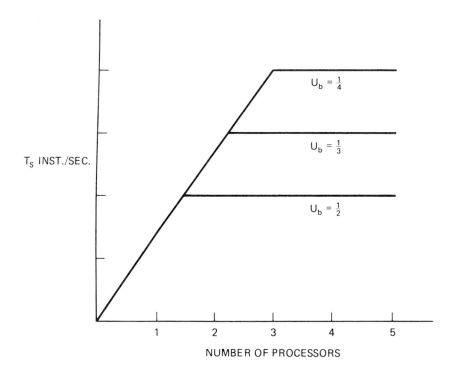

*Figure 4.16: Throughput for a Common Bus*

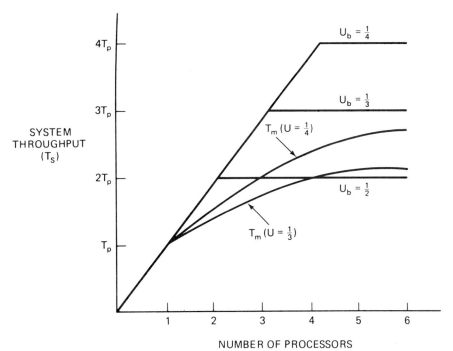

Figure 4.17: System Throughput — Worst Case

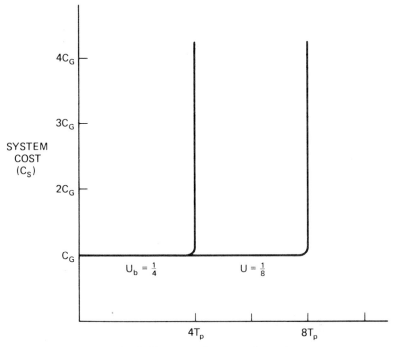

Figure 4.18: Systems Cost vs Throughput

throughput as observed earlier is not necessarily a good indicator of the useful work being done by the whole system. Indeed, if the overhead required to synchronize and coordinate the system increases disproportionately as processors are added, the useful work may decrease. On the other hand an additional processor may allow a more effective partitioning of tasks with a reduction of overhead and the useful work may increase even if instruction throughput does not.

The shared resources such as memory and peripherals usually cost several multiples of the processor cost. Indeed more effective use of such bus resources could be argued as sufficient reason for adding processors. If the global resources cost $C_g$ and each processor $C_p$ then the system cost is $C_s = C_g + NC_p$. Processors are usually a small fraction of the cost of such items as memory or I/O devices. Such costs are changing rapidly with time. If $C_g = 50C_p$ for example, then the cost of throughput can be plotted using the data from Figure 4.16. This is shown in Figure 4.18. For this example: for $U_b = 1/4$

$$C_s = C_p [1 + 50N] \text{ for } 0 \leq T \leq 4 T_p$$

$$= \infty \quad \text{ for } T_s > 4 T_p$$

Costs increase to infinity since no further throughput is obtainable.

If systems cost per instruction were now plotted against the number of processors, it is obvious that a rapidly declining curve would be obtained which would at the $N = 1/U_b$ rise rapidly. It is possible to compute the minimum cost per instruction and choose the appropriate number of processors.

The preceding discussion and analysis leads to several more important considerations:

1. effect of private memory

   The addition of private memory reduces the apparent value of $U_b$ and hence the critical points in all the associated curves. The utilization of private memory requires partitioning of tasks so that system throughput approaches $NT_p$. The attainment of this goal is a measure of the success of the task partitioning and the synchronization and control procedures.

2. Bus Utilization and Instructions

   Processor design for multiple processor applications should lower $U_b$. This implies that minimal access to common memory during a complete instruction cycle should be required. Processors which have a large number of internal registers with supporting operations should require less bus access for a large variety of bench mark programs.

### 4.4.3    Arbiters

Commercially available multiple processor boards generally provide facilities for creating distributed systems as we have defined them. Each board has room for private memory and usually both serial and parallel output ports for peripherals. The memory map as viewed by the processor can be partitioned by switches to a private and a public port. References to public memory are trapped and become a request for bus access to the arbiter. A similar partition is made of the input/output address space. Following a grant the processor is connected to the common bus as a bus master. This connection is typically for a the duration of memory access only. In some cases a bus lock/unlock facility is provided. The lock command maintains control of the bus (after grant) until it is released by unlock.

### 4.4.4    Resource Protection

The problem of resource protection is a general problem in designing concurrent systems. A test and set flag was explored in Chapter 1 as a logical means of creating a critical region of code. Equally such flags can be used by an operating system to designate peripheral resources on the common bus as available or not. Indeed the kernel of the operating system is a common resource to all processes and monitors and as was stated in Chapter 4 hardware arbiters must finally process simultaneous requests. Microprocessors are beginning to appear on the market with an indivisible test and set instruction and thus software flags are easy to implement.

A hardware test and set flag can be implemented as a common bus peripheral as shown in Figure 4.19. Such a flag could be memory mapped or treated as an input/output port. On a read instruction the contents of the flip-flop are transferred to the processor and the flip-flop set to 1. On a write the flip-flop is set to 0. Indivisibility is assured by the arbiter provided the instruction (Read or Write) is indivisible. This is usually time for an input/output instruction and since memory space may be scarcer than peripheral space, the TAS was designed as peripheral.

The special board contains four flags which are mapped as I/O addresses (rather than memory). From the assembly language perspective an IN TAS performs the following indivisible operation:

```
IN TAS=
        [ACC := FLAG
         IF FLAG = 0, THEN FLAG := 1]
```

similarly

```
OUT TAS =
        [FLAG := 0]
```

The *IF THEN* application is performed by hardwired logic as shown in Figure 4.19.

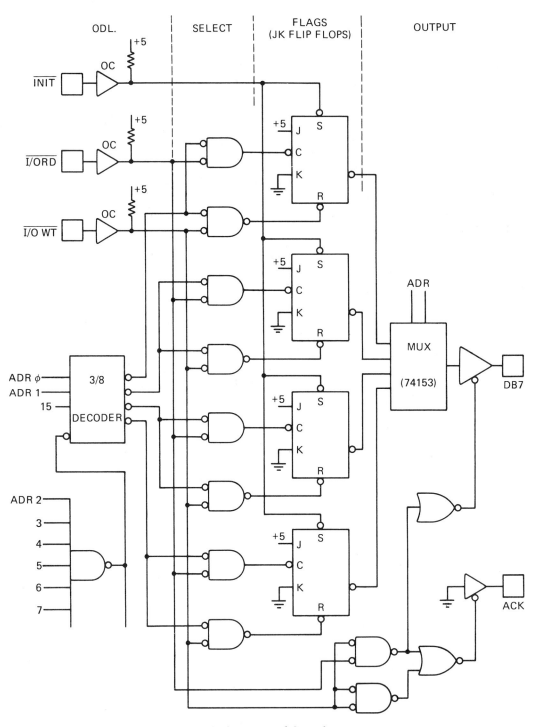

*Figure 4.19: Test and Set Flags*

The design of a TAS would require attention to the details of the particular bus protocol and timing signals. Figure 4.19 shows an implementation for the INTEL 8080. The four flags are coded as I/O addresses 00, 01, 10, 11. The hardware is divided into four logical functions:

1.  Decode

    A 3/8 decoder examines the low byte of the address bus for the appropriate four addresses.

2.  Select

    Two AND gates per flag are used to gate I/O RD or WT to the appropriate flag. I/O RD clocks a one into the flag on the falling edge. I/O WT causes a reset.

3.  Flag

    The flags are set or reset by I/O RD and I/O WT. Note the change in state on RD takes place on the falling edge and therefore the state of the flag is read before it is set.

4.  Output

    A 4/1 multiplexer selects one of the flags which is gated to the data bus. The bus protocol requires an acknowedge signal which is created as shown.

    The design of such flags requires the obvious precautions to insure that the operation is indivisible. The details of utilizing the bus signals and the characteristics of standard logic components are straightforward.

## 4.5 Processors for Multiprocessing

Early microprocessors of the I8080, M6800 or Z80 class were not designed for multiple processor applications. The I8080 fortuitously has the extra hold state during instruction fetch which was designed to accommodate slow memories. This state proved ideal for creating a multiple processor board (the SBC 80/20). Potential delays in the arbiter response for bus access could be made to appear as a slow memory as explained in Section 4.3. There are indications that equivalent gate density on a processor chip will achieve 5 to 10-fold increases by 1985. This creates the chip real-estate for impressive increases in capacity. Part of this capacity could be used for enhancing multiprocessor architecture depending on perceived market demands. Clearly the availability of higher powered microprocessors tends to ameliorate the requirements for utilizing more than one processor; on the other hand the same increase in capacity leads to the potential for awesome performance/cost ratios when several are coupled to perform a task.

Many of the logical features required for multiprocessing are also useful in all applications and hence a general improvement will probably occur by default. In this section capabilities which would enhance utilization in a multiprocessor application will be discussed. Generally of course any improvements which yield an efficient assembly language realization from a high level language is a positive factor.

The capabilities for implementing the logical constructs inherent in concurrent systems require efficient mechanisms for establishing mutual exclusion and for efficient context switching. In addition hardware capabilities must be available for interrupt servicing, bus and arbiter interfacing, and hopefully for direct interprocessor synchronization.

Mechanisms for establishing mutual exclusion involve some form of semaphore operation. This involves both a TAS operation, manipulating a counter and queueing. Newer processors include an indivisible TAS instruction which can address and modify a Boolean flag in memory. The rest of the semaphore operation must be implemented in available instructions. The addition of increment/decrement counter and conditional branch would be of additional value. Finally efficient queue manipulation affects both this operation and others and will be discussed as part of context switching.

Context switching is a general term to describe the process state transitions required to implement the required system. Such switching requires extensive queueing and stack operations.

The creation of data structures for PCBs and associated queues requires good addressing modes for efficient implementation. Thus a full range of indirect, indexed modes are clearly desirable. Stack manipulations are involved in many aspects of multiprocessing.

The complete save/restore of all processor registers during ISRs (to be discussed later) and each process state transition is perhaps the single factor contributing to the overhead of concurrency. Thus many CPU registers which contribute to high speed internal operations on the one hand tend to create slow stacking response times during context switches. Architectural innovations which create memory registers as high speed working registers are clearly advantageous. Stack instructions which quickly save and restore the processor control registers (PC, STATUS, etc.) tend to minimize switching times. The implementation of reentrant subroutine calls is also an important feature of a system (discussed more fully in later chapters) which deserve special attention in the stack manipulative structure of the processor.

The most outstanding unique feature of a multiple processor hardware configuration is the requirement to submit to arbitration its demands for global system resources. In the common bus system discussed earlier, the bus was the global resource. Access to the bus allowed access to other global resources such as memory or peripherals. Memory maps could be created and suitable decoders set up to trap off-board references which become an arbiter request. Arbitration implies that on simultaneous requests, one or more processors must wait. This implies that the processor must have a WAIT state or some mechanism must be found for suspending the processor in a queue. Complex arbiter chips are appearing which promise to allow more flexible allocation of the partition among processors of global resources. Also most useful would be direct processor/processor synchronizing which would facilitate the partitioning of an access graph among processors in a more arbitrary way.

Interrupts create the logical equivalent of an unpredictable asynchronous subroutine call. Such calls have two major requirements; they should not affect the algorithm they interrupt, except for a time delay which should be as short as possible.

The guarantee of the continuing validity of the systems algorithm in the presence of asynchronous interrupts has and probably will continue to be a difficult responsibility of the systems designer and implementer regardless of the processor being used. Minimizing the overhead of responding to interrupts involves efficient context switching (discussed earlier) and effective ways of identifying the interrupt source and establishing the location of the appropriate service code. Vectored interrupts which provide pointers to service code are widely available as auxiliary chips in most processors.

In addition to these requirements it is usually desirable to establish a system of priorities among different devices. This feature is also available in priority interrupt control chips. Operationally it is often desirable to have effective means of dynamically altering the priorities or masking interrupts from various sources while servicing a particular source. Purely software implementations (in association with an interrupt chip) tend to increase response time. Multiple levels of interrupts as a hardware feature of a processor chip creates a pin-out requirement which often limits the interrupt levels to one. Processors with multiple levels (including a non-maskable level) increase the flexibility of implementing interrupt algorithms while reducing the overhead (which tends to reduce response time).

The rapid increase in equivalent gate density of all chips indicates that higher performance processor and auxiliary chips will continue to become available in response to market demands. The maturing perspective of the divergent application areas should insure an increasing awareness of implementation requirements both on hardware architectures and instruction sets. These will in general benefit the implementation of multiple processor systems. The manufacturer's perception of the magnitude of the multiple processor applications market will be a major factor in the inclusion of specific features to accommodate this segment of their projected market.

## 4.6  References

Several multiple processor companies are comparisons in the following:

S.H. FULLER, "Price/Performance Comparison of C.mmP and the PDP-110", 3rd Annual Symposium on Computer Architecture, 1976, pp. 198-202

A. ROSENBLATT, "How GI Captured the OTB Stakes", Electronics, Mar. 1, 1973, pp. 66-67

H.E. CONNELL, "A Multi-Minicomputer Network for Optical Moving Target Indication", Digest of Papers, Compcon Fall, 1976, pp. 233-235

R. McGill and J. Steinhoff, "A Multi-microprocessor Approach to Numerical Anmalysis: An Application to Gaming Problems", 3rd Annual Symposium on Computer Architecture, 1976, pp. 46-51

J.E. Wirshing, "Computers of the 1980's — Is It a Network of Microcomputers?" Digest of Papers, Compcon Fall 1975, pp. 23-26

## Architectural classifications have been proposed in:

M.J. Flynn, "Very High Speed Computing Systems", Proc. IEEE, Vol. 54, No. 12, Dec. 1966

L.C. Higbie, "Super Computer Architecture", Computer, Dec. 1973, pp. 48-58

## Interconnection topologies are well discussed in:

G.A. Anderson and E.D. Jensen, "Computer Interconnection Structures: Taxonomy Characteristics and Examples". ACM Computing Surveys, Vol. 7, No. 4, Dec. 1975, pp. 197-214

C.V. Ramamoorthy and T. Krishnarao, "The Design Issues in Distributed Computer Systems", Infotech. State of the Art Reports on Distributed Systems, 1976. Maidenhead, Birks, England

## A source reference paper on bus structures and protocols:

K.J. Thurber, et al, "A Systematic Approach to the Design of Digital Bussing Structures", Proc. IEEE Fall Joint Computer Conference, 1972, pp. 719-740 (contains also an extensive bibliography)

## An excellent text with extensive coverage of the hardware features of multiple processors is:

C. Weitzman, "Distributed Micro/Mini Computer Systems," Prentice-Hall, Englewood Cliffs, N.J.

Chapter **5**

# The Kernel

## 5.0  Introduction

The kernel is a critical region with special responsibilities in the implementation of concurrent systems. Within the kernel a set of procedures are executed which coordinate the concurrency of the remaining software structures. The kernel is often referred to as a virtual machine and the procedures called "primitives". The primitives appear to the higher levels of software as single monolithic commands — much as the assembly language commands of the processor on which they run.

Two types of kernels are often used in concurrent systems. The first is a synchronizing kernel which provides only the mechanisms for implementing the state transitions of the processes. And the second is a larger message passing kernel which becomes directly involved in the physical movement of data. The synchronizing kernel is the simplest and generally provides the highest performance.

Because it controls all state transitions of all processes, the kernel is clearly a potential bottleneck. It is the task of the designer (and implementer) of the kernel to create the virtual machine in a highly efficient fashion. It must provide not only the required primitives but accommodate the hardware architecture and introduce minimal delays due to its presence. As we shall discuss in later chapters, the kernel is often the place where top-down design requirements and bottom-up imposed constraints must be resolved. The presentation here emphasizes synchronizing kernels. Future systems of loosely coupled microprocessors may demand versions of message passing kernels; however, their design is not as well understood at this time.

The material is divided into two parts. Section 5.1 provides an overview of the logical structure of a kernel. Each of its components is explored in some generality at the level of pseudo code. In Section 5.2 an example of a real multiple processor synchronizing kernel is presented. This kernel was designed and implemented for an Intel SBC 80/20 system. Example code at the assembly language level is used to insure a complete understanding of implementation details.

Further kernel issues are postponed until Chapters 7, 8 and 9. Chapter 7 reviews kernel requirements for the example packet switching system. Chapter 8 discusses

CHLL interface requirements for both PL/M and PASCAL. And Chapter 9 addresses efficiency and testing/debugging issues, among others.

## 5.1    The Logical Structure of Kernels

### 5.1.0    Introduction

The onion-skin view of a concurrent system proposed in Chapter 1 (Figure 1.8) can be expanded in its layers as shown in Figure 5.1. This figure shows only the software structures discussed thus far. The inner ring contains the basic semaphore structures used for synchronization as discussed in Chapters 1 and 2.

The outer ring contains the processes which were discussed in Chapter 2. Next there is the monitor as discussed in Chapter 3. The monitor ring is subdivided into two segments in which the monitor scheduling mechanisms are isolated. The kernel layer can be drawn so as to include these mechanisms or to include only the basic semaphore operations. The choice affects all measures of kernel performance.

The monitor scheduling mechanisms discussed in Chapter 3 could be incorporated into the kernel and called by the monitor. On the other hand the kernel boundary could be moved inward to include only basic semaphore operations, leaving the scheduling to the monitor designers. The latter type of kernel will be discussed in Section 5.2. Indeed, as will be shown in Chapter 9, the kernel ring can be made even smaller by eliminating queueing altogether. The trade-offs are system dependent. Clearly, if kernel primitives are limited to basic semaphore operations, the  kernel will be smaller and more flexible than if monitor scheduling mechanisms are included. If the system is a prototype one or one with adaptability requirements, it may prove wisest to leave monitor scheduling outside the kernel. However, speed of monitor scheduling may be enhanced by including it in the kernel. Such are the usual design trade-offs facing any systems designer.

Regardless of the design decisions with respect to the kernel boundary it is evident that the kernel contains four main parts:

— The CHLL interface

— The kernel main

— The Procedures

— Data structures

In Section 5.1.1 a more concise statement of the general requirements of a kernel are proposed. In the subsequent four sections we shall explore these functions in more detail. Finally interrupt handling decisions must be made as well and we shall look at alternatives in Section 5.1.6.

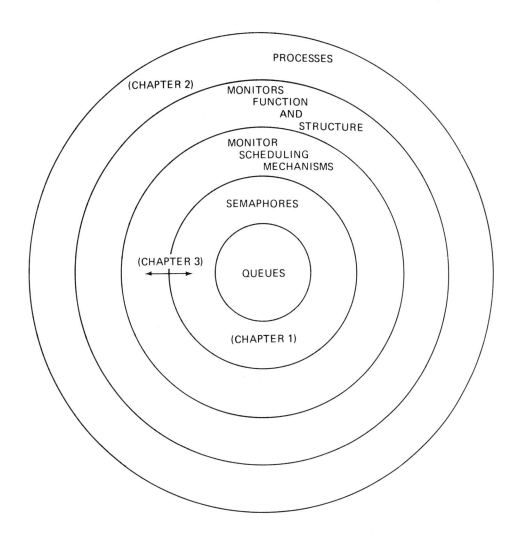

*Figure 5.1: Expanded Onion-Skin View of Multipleprocessor Software*

### 5.1.1  General Requirements

The kernel is a critical region which controls systems functions associated with concurrency. Because of its central role, it can easily become a bottleneck, seriously limiting throughput. Because it maintains concurrency of system processes and monitors as well as accommodating the hardware configuration, the kernel usually has a set of requirements more stringent than other critical regions in the system. These can be enumerated under the general categories of access, execution and structure.

In general kernel specifications must include a sufficient array of primitives and associated data structures to support the system functions. In addition real time constraints usually require high speed execution and for good measure the kernel should occupy as little memory space as possible. Finally, the kernel must often be able to accommodate variations in the hardware family which might be used in a specific application. More specifically these requirements imply the following:

Access:

Access to the kernel is the same logical problem as access to a monitor. However, the arbitration problem which was postponed in discussing monitor gate access must now be faced. A monitor gate consisted of a semaphore with indivisibility guaranteed by the kernel. The kernel arbitrated simultaneous requests to execute the semaphore code.

Kernel calls differ from monitor calls in two ways. First the arbitration problem must now be solved. And this is usually done by a hardware device similar to the arbiters used to allocate a common bus as described in Chapter 4. Secondly, the semaphore function is altered to eliminate the associated queue. This implies of course that calling processes are forced into a busy wait condition. This is normally an optimal solution since the kernel code usually imposes only short delays.

If a busy-wait is not desirable (for example because of, say, bus traffic congestion) then the processor must be suspended and its request noted until its turn. Once again this is a hardware function built into an arbiter.

Execution:

Demands for high speed execution of kernel primitives almost invariably require highly optimized code, which in turn suggests assembly language implementations. This brings about the usual range of problems of design, documentation, and debugging. In addition, the kernel must interface to the logical view presented to higher levels of software, in particular parameter passing and returning.

Structure:

The internal structure of the kernel is logically separable into three parts — the main, the procedures and the data structures. The main is entered after passing the kernel gate. Its functions can include interpreting the parameters passed from the calling process, as well as a variety of validations and security or operational logging. Its final task is to call the appropriate internal procedure which implements the primitive. The procedures and data structures must be designed to execute the desired system functions and to maintain in the data structures all the required records for implementing the process state transitions and, if applicable, the monitor scheduling mechanisms.

The kernel can be viewed as a procedure call by a process or monitor in which primitives are defined by parameters, e.g.:

```
                PROCEDURE:  KERNEL (PARAMETERS)

/*ENTER*/
        DISABLE PROCESSOR INTERRUPTS
        SAVE PROCESSOR REGISTER STATE IN STACK
        REQUEST KERNEL (ADDRESS ARBITER)
/*MAIN*/
        SAVE PROCESSOR STACK POINTER
        VALIDATE PARAMETERS
        CALL PROCEDURE
/*EXIT*/
        RELEASE KERNEL (ADDRESS ARBITER)
        RESTORE PROCESSOR REGISTERS
        ENABLE INTERRUPTS
        RETURN
```

The design and implementation of each aspect of this procedure is obviously most crucial if the usually opposing requirements of flexibility, small size and high speed, are to be accommodated. The next four sections will explore the components of the kernel in more detail.

## 5.1.2   CHLL Interface

The CHLL interface includes the entrance and exit portions of the kernel procedure. System requirements usually require that kernel calls appear as a standard subroutine or procedure with parameter passing conventions compatible with the programming language. The scheduling or other functions of the kernel should normally be transparent to the caller.

The entry routines as shown in the latter part of the previous section include three parts:

— Disable interrupts

— Save processor register state

— Request kernel

This procedure is similar in some respects to a uniprocessor interrupt service routine. The processor interrupts are disabled under software control since the kernel is an interrupt protected region and the register saving should not itself be interrupted.

The processor register state is saved since the kernel normally executes on the calling processor. If this were not the case (e.g., a separate kernel processor) the step might be eliminated. Since the kernel needs access to the parameters being passed which are normally stored in machine registers, it is necessary to store the registers in a location accessible to the kernel processor.

The kernel request implies a call to the kernel arbiter which is normally a hardware device. This may require a separate hardware arbiter accessible by all processors. Or as shown in Section 5.2, the existing common bus arbiter is used to control sequential access to a common hardware test and set flag.

The main distinguishing feature of arbiters is the mechanism used to force waits on a processor if the kernel is busy. As with monitors a busy wait or a queued wait is possible. In monitor implementations a busy wait was undesirable because of the long delays which were possible during monitor execution. Busy waits by several calling processes could create unacceptable bus traffic and contribute to performance degradation. For the kernel, however, this may not be as serious since kernel execution is comparatively short and not subject to unpredictable delays. Busy waits are also often the simplest to implement. Such a scheme will be shown in Section 5.2.

A queued wait implies a more complex arbiter and also some mechanism for suspending the processor that is waiting. This may or myay not be possible depending on the control pins (e.g., the I8080 has a Hold line which makes it possible). If such hardware features are not available, a software loop with an interrupt generated by the arbiter is a possible solution. Queued waits for processors have not been used widely for these reasons.

Exit from the kernel was shown to contain four parts:

— Release the kernel

— Restore processor registers

— Enable interrupts

— Return

Releasing the kernel involves a call to the arbiter. In the case of a hardware test and set flag, this implies a reset of the flag. The processor state must be restored, interrupts enabled and a standard return executed. This procedure is very similar to a return from an interrupt service routine.

We note here that the kernel procedures deal with the register contents of the processor which are stored in a location known to the kernel. The parameters or pointers to parameters are stored here and return information is left in this data structure.

### 5.1.3   Kernel Main

For our purposes the kernel main is defined as that portion of the kernel immediately following the entry and preceeding the primitives. In the simplest case this is just a

CASE statement however, further functions are often desirable, particularly call and parameter validation, and journalling.

It is possible to include in the kernel a series of validation checks on incoming calls, particularly during system creation or modification, or perhaps in a multi-user environment. The parameter passed to the kernel can be checked for out-of-range values for example, and if limitations exist on useable primitives for some processors, these can also be enforced.

Various journalling procedures can also be incorporated. During system testing or modification a debug flag can be set which causes a record to be maintained of significant steps through the kernel. Such a journal can be used for debugging or perhaps for quantifying the usage of various kernel features.

### 5.1.4 Procedures and Data Structures

Following the main portion of the kernel the procedures implement the primitives. At this point the designer and programmer are on familiar ground. The overall design can proceed by defining the functions to be performed by each primitive and the data structures needed to support these functions.

The primary responsibility of the kernel is to support the concurrency implied by the state graphs of processes and, if applicable, the scheduling requirements of the monitors. Efficient routines are therefore desirable for testing software flags and for entering into and obtaining process identifiers from queues. In addition, efficient search and update routines for process control blocks are required.

The content of the process control blocks and the queueing structure is application dependent. The process control block must contain sufficient identification and status information that a process can be safely queued and subsequently restarted. This involves the register state of the processor running the process when it was queued at a minimum, but may involve other system constraints such as the process priority, starting address, stack pointer and possibly supporting processors. The content is normally arrived at in a heuristic way by "walking" the process through all possible sequences of queues. Typically the following may be found necessary:

— Stacks: One for each process and one for the kernel
— Semaphores: One for each state and condition
— Process Control Blocks: One for each process, which usually contains the stack pointer, process index, priority, blocked state, process or index
— Processor Information Table: Identifies which processors can run which process and the currently running process
— Ready-to-Run Queues: One for each processor

Process stacks do not, of course, belong to the kernel; they are required by the processes. Process control blocks and semaphores may belong to the kernel and be identified by index or may be supplied by the process and identified by address. These alternatives are discussed further in Chapters 8 and 9. The examples of this chapter assume they belong to the kernel and are identified by index. The remaining structures belong exclusively to the kernel.

### 5.1.5    Interrupts, Interrupt Service Routines (ISRs) and I/O

The primary purpose of this section is to demonstrate the difficulties and trade-offs which must be considered when interrupts must be accommodated. Many of the classical approaches are based on a single processor implementation. We will briefly review them and suggest an alternative.

Because kernel primitives should be both "universal" and efficient, short term scheduling policies for kernel calls are likely to consist of simple First-In-First-Out priority algorithms. If, however, a monitor protects access to a shared I/0 device, a disk for example, then rather than having device requests enter a First- In-First-Out queue, it is possible to have them join a more complex but more appropriate queueing structure.

If more than one system process is allowed to access directly a single shared device

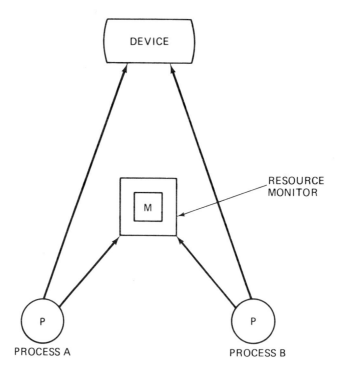

*Figure 5.2: Use of a Resource Monitor to Protect Shared Device Access*

(see Figure 5.2), then the procedure for a process to perform I/0 is as follows:

1. Call the resource monitor to request use of the device.

2. Ask kernel to perform I/0 when authorized to do so.

3. Call the resource monitor to release the device when finished.

A process must request use of the device by calling the resource monitor which may or may not, depending on the number of other requests at the time, delay the calling process in a user specified medium term scheduling queue. In Concurrent Pascal for example, the kernel itself is responsible for all direct communications with the various I/0 devices. Once a process has been given permission by a resource monitor to use a particular device, it may originate I/0 by calling a standard kernel I/0 routine.

The concept of a resource monitor thus ensures proper synchronization between all processes which simultaneously request access to a shared device. The overhead associated with I/0 in such a system is however considerable. Timing diagrams for a typical I/0 sequence are given in Figure 5.3.

Two major efficiency problems may immediately be recognized:

1. At least five kernel calls are required to initiate I/0 if the device is busy at the time of the request. (ENTER MONITOR, DELAY, CONTINUE, EXIT MONITOR and START I/0). More are required if the monitor scheduling procedures are not part of the kernel.
2. The device utilization ratio is considerably degraded by the numerous kernel calls which are required between the termination of one I/0 request and the start of another request.

One possible way of reducing the overhead, and thus of improving overall efficiency, consists of implementing the medium-term scheduling policies as integral parts of the kernel. A "START I/0" kernel call could then automatically queue requests in an appropriate scheduling queue, without needing to refer to a resource monitor. The resulting timing diagrams for the synchronization of processes wishing simultaneous access to a shared device would then be as shown in Figure 5.4. From them, it may be seen that this solution has resulted in the elimination of most of the previously required kernel calls, and that the device idle time between I/0 operations has been reduced considerably. The resulting advantages are thus significant. Unfortunately, they are obtained at the expense of added complexity of the kernel which now performs all medium term scheduling.

A fundamentally different approach to I/0 consists of assigning a specialized I/0 process to each I/0 device. These processes would then carry total responsibility for all direct communications with their associated device, through device handlers separate from the kernel as discussed in Chapter 2. Other processes could request I/0 by

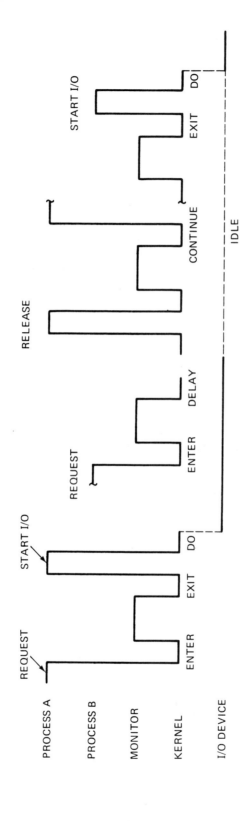

*Figure 5.3: CHLL Scheduling*

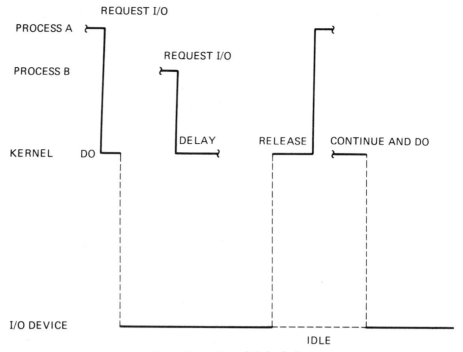

*Figure 5.4: Kernel Scheduling*

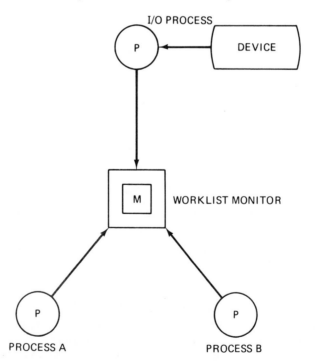

*Figure 5.5: Use of a Worklist Monitor to Buffer I/O Requests*

depositing proper requests, with all necessary parameters, in a worklist monitor which virtualizes the device. The resulting system configuration would then be as shown in Figure 5.5. While waiting for their request to be serviced, processes could either be delayed inside the worklist monitor, or delay themselves by waiting on a semaphore, which the I/0 process would then signal as soon as it was finished servicing the request. Several advantages may be associated with this type of solution to the I/0 problem:

1.  The device utilization ratio improves, since fewer kernel calls are required between the servicing of two successive requests.

2.  An overall simpler kernel which is no longer responsible for I/0, results.

3.  Special processors could be dedicated to I/0 processes, thereby eliminating the need for total interconnection between all peripherals and all system processors.

The difficulties with interrupts are their random occurrence, and with ISRs, their real time response requirements. Both of these can be accommodated if the design principles suggested in Chapter 2 are adhered to. It was pointed out that kernel code is executed with interrupts disabled. Careful timing estimates must be made to insure that sufficient time is available to respond to interrupts. When an interrupt is acknowledged the ISR becomes a critical region under the protection of the automatic interrupt disable. Thus ISR calls to the kernel can utilize special entry mechanisms. An example will be shown in Section 5.2.1.

A producer and consumer process can easily exchange data pointers if suitable kernel primitives are available, i.e.,

ATTACH (buffer queue #N, address of buffer)
SIGNAL (semaphore #M)

The consumer can receive the buffer by calling the primitives

WAIT (semaphore #M)
DETACH (buffer queue #N, address of buffer)

The consumer can then return the empty buffer to the producer by the same mechanism. Clearly a process and ISR could exchange buffers this way provided the ISR never performs a WAIT. Thus ISR to process buffer transfers are straightforward. In the reverse direction the process can attach a buffer but not signal a semaphore. The ISR detaches the buffer. If it is empty the kernel returns a zero.

This solution can cause special problems depending on the device interface. Consider for example a UART feeding a CRT display. Such a device will transmit a character, then generate an interrupt indicating a request for a new character. If the ISR returns from an interrupt without dispatching the next character (i.e. the buffer was empty) then no further interrupts will occur and the CRT appears to have failed. This possibility is not hard to accommodate as was discussed and illustrated in Chapter 2.

### 5.1.6   Summary of Design Issues

It is the logical function of the kernel to transform the lower levels into a virtual machine which provides the facilities for implementing the concurrency requirements of the system. In our examples the kernel provided the initialization of the system and the primitives for process/monitor/process interaction. An appropriate set of such primitives is clearly a first consideration in the specification of the kernel. In addition to the specific primitives there is usually an execution time limit which results from system response constraints. The definition of primitives and their functions are often an iterative design problem if both constraints are to be met.

Response constraints also force the consideration of appropriate entry and exit mechanisms. High speed arbiters and test and set facilities are usually required to minimize delay at this point.

Advantages in efficiency, modularity and simplicity are gained by keeping device interrupt handling separate from the kernel. As part of the onion-skin view of Figure 5.1, ISRs fit into the process layer as degenerate processes and device handlers fit into the monitor layer as degenerate monitors. In both cases constraints exist on the interfaces to lower layers, as has been explained both in Section 5.1.5 and in Chapter 2. And in both cases, access is made directly to the innermost hardware layer (which is not explicitly shown in Figure 5.1).

The kernel is obviously constrained by the hardware architecture which itself is often constrained by system specifications. In Chapter 7 an "edge-in" design procedure accounts for this pincer approach to the kernel. The kernel designer is often in the unenviable position of attempting to solve all the problems left over from other phases of the system.

### 5.2   An Example Kernel

### 5.2.0   General Description and Organization

In this section a monolithic software kernel designed and implemented for a multiple microprocessor system will be described in detail. By this example, the realities of a microprocessor environment will be introduced and perhaps the detail will help to fasten the conceptual discussions presented thus far to a real environment.

The kernel discussed here was designed to

— support a real time multiprocessing operating system written in PL/M* (for Intel 8080 microprocessors)

— provide communications between processes and ISRs

— run on an Intel SBC 80/20 (common bus) system augmented with a hardware TAS flag to control kernel access

*PL/M is subset of PL-1; supported by Intel Corp.

The design goals were to produce a fast simple and reliable kernel which would support four or more processors. It is coded in assembler with parameter passing conventions compatible with PL/M.

A design decision that processors are to be functionally specialized restricts a process to run on only one processor designated at initialization, although limited facility is available to change this as will be shown. The RTR queue of each processor consists of a foreground and background FIFO queue and each process is permanently assigned a fixed priority.

The kernel primitives support process switching. Primitives are also provided to initialize the kernel and to pass buffers. The overall software organization is shown in access graph form in Figure 5.6 while the internal kernel organization is shown in Figure 5.7. Our purpose in the following is to examine each component in detail. Before proceeding we note the hardware architecture is shown in Figure 5.11. The common bus structure is augmented by private memory in an attempt to reduce bus traffic. A hardware test and set flag (discussed in Chapter 4) is used to enforce mutual exclusion on the execution of the kernel.

The arrows in the access graph drawn from those system components outside the kernel to the kernel data area indicate access by processes and ISRs to the process stacks (which are included here as part of the kernel data). Additionally they indicate access by processes to a special variable called NAME described below.

The kernel data includes, in addition to those items mentioned in Section 5.1.4, the following special items for each processor:

— a memory location called NAME containing the index of the currently running process;

— an I/O port address called ME which is permanently wired to return the index of the particular processor when addressed by an IN instruction.

These items are used as follows:

— NAME is updated by the kernel every time a new process is dispatched to run and is used by the monitor scheduling primitives to obtain the index of the currently running process for deposit in a high level monitor queue (see Chapter 3, Figure 3.18);

— ME is used by the kernel to identify which processor it is running on, so that it may select the correct RTR queue.

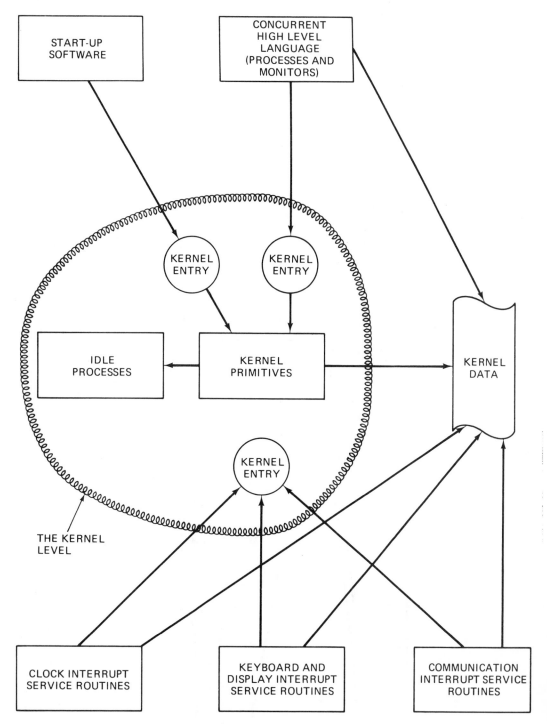

*Figure 5.6: Overall Software Organization*

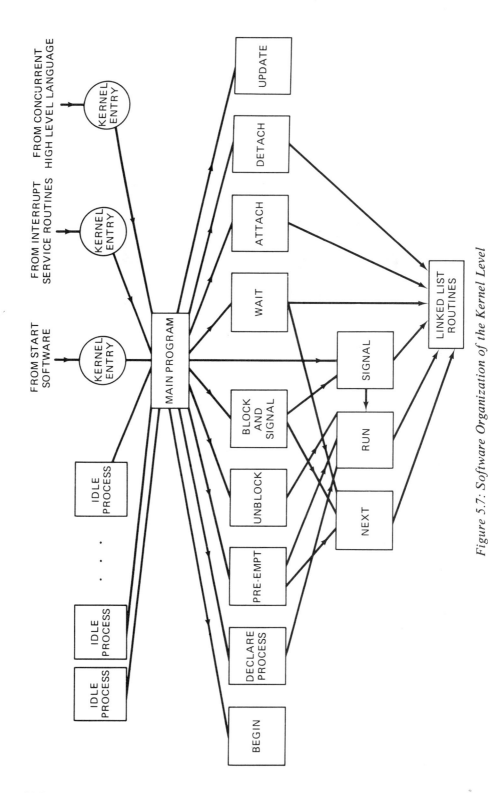

*Figure 5.7: Software Organization of the Kernel Level*

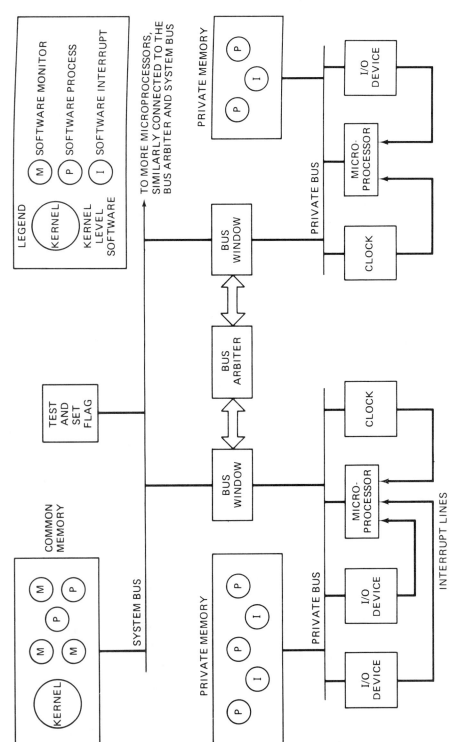

Figure 5.8: Hardware Configuration and Location of Software

### 5.2.1   Kernel Entry/Exit

Normal procedure calls in PL/M have a pre-defined parameter passing convention for the Intel 8080 architecture as follows: The register pair BC contains the first parameter (of type address); the register pair DE contains the address of the buffer holding the remaining parameters.

All kernel calls identify a primitive with a code in register B, as follows:

| | |
|---|---|
| 0 — BEGIN | 5 — BLOCK & SIGNAL |
| 1 — DECLARE PROCESS | 6 — UNBLOCK |
| 2 — PREEMPT | 7 — ATTACH |
| 3 — WAIT | 8 — DETACH |
| 4 — SIGNAL | |

Figure 5.9 shows a typical call from a process (in PL/M) to wait on semaphore 5. As part of the code, Kernel is declared literally CHLL and WAIT as 03$05H. Kernel is an external procedure as shown. The call loads B and C, then jumps to CHLL which is the appropriate entry routine. The flow of the call through the kernel is shown in Figure 5.10. The corresponding I8080 code is shown in Figure 5.11.

Note that interrupts are disabled immediately. If the TAS indicates the kernel is busy, the interrupts are enabled briefly during each loop on TAS, to minimize the possibility of missing interrupts. The accumulator must be saved because it is used by the TAS operation. The remaining registers do not necessarily have to be saved at this time.

The kernel main is shown in Figure 5.12. The stack pointer of the calling process is saved, the kernel stack substituted, and a JUMP is performed to the primitive designated in register B. Note in the code (Figure 5.13) that DO CASE is performed to find the required primitive. This is somewhat difficult to implement simply due to the limitations of I8080 code.

```
        DECLARE KERNEL LITERALLY 'CHLL';
        DECLARE WAIT LITERALLY '03$05H';
KERNEL:
        PROCEDURE (BC$REGISTERS, DE$REGISTERS) EXTERNAL;
        DECLARE (BC$REGISTERS, DE$REGISTERS) ADDRESS;
        END KERNEL;
/*PROCESS*/

        ...
        CALL KERNEL (WAIT, 0);
          ...
        END;
```

*Figure 5.9: Process CALL — Wait on Semaphore*

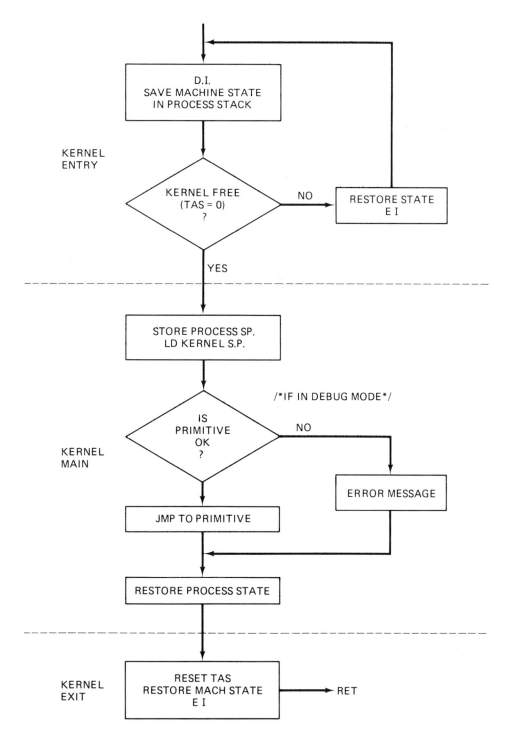

*Figure 5.10: Flow of Control: Kernel Entry and Exit*

127

```
; MACROS
TEST MACRO ABC        ; THIS MACRO DEFINITION ALLOWS TEST TO BE USED
     IN 03H           ; TO TEST THE TAS FLAG.
     ENDM             ; ABC IS A DUMMY PARAMETER REQUIRED BY THE COMPILER
RESET MACRO ABC
     OUT 03H          ; H INDICATES HEXADECIMAL NUMBER
     ENDM
CHLL:  DI             ; DISABLE; INTERRUPTS
                      ; SAVE PROCESSOR STATE IN MACHINE STACK
     PUSH PSW
     PUSH B           ;
     PUSH D           ;
     PUSH H           ; SAVE PROCESSOR STATE IN MACHINE STACK
     TEST OOD         ; WAIT ON TAS FLAG, D IS A DUMMY PARAMETER
     DCR A            ; DECREMENT ACC
     JM CHH 1         ; JMP TO CHH1 IF NOT SET

     POP H            ; IF BUSY RESTORE MACHINE STATE
     POP D            ; CHECK INTERRUPTS AND TRY AGAIN
     POP B            ; AGAIN
     POP PSW          ;
     EI               ;
     NOP              ; NO OPERATION, REQUIRED TO WAIT FOR INTERRUPT ENABLE
     JMP CHLL

CHH1:
     CALL KERN        ; ENTER MAIN KERNEL PROGRAM
CHH2:
     ;KERNEL EXIT FOLLOWS

     RESET D          ; RETURN RESET TAS
     POP H            ; RESTORE MACHINE STATE
     POP D            ; ENABLE INTERRUPTS & RETURN
     POP B            ;
     POP PSW          ;
     EI
     RET
```

*Figure 5.11: Kernel Entry and Exit: I8080 Code.*

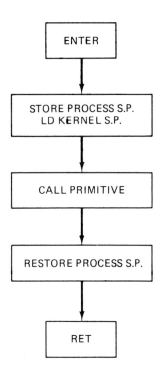

*Figure 5.12: Flow of Control: Kernel Main*

```
        /* Store Machine, Stack Pointer and Get Kernel Stack
Printer */

    KERN:    LXI   H,OD        ;   CLEAR HL
             DAD   SP          ;   HL←SP
             SHLD  STACK       ;   STORE STACK
             LXI   SP,KRNSP    ;

    /* AT THIS POINT THE PRIMITIVE BYTE B COULD BE */
    /* EXAMINED FOR VALUE    THIS IS CONSIDERED */
    /* IN SEC. 4.4 */

             LXI   H, KER 1    ; (SET UP A DO CASE
             PUSH B            ;  ON B TO GET TO PRIMITIVE)
             MOV C, B          ;  C←B
             MVI B, OD         ;  B←O
             DAD B             ; (EACH JUMP IN KER1
             DAD B             ;  TAKES 3 BYTES)
             DAD B             ;
             POP B             ;  RESTORE BC
             PCHL              ;  PC←HL

    KER 1:   JMP   BEGIN
             JMP   DECLARE
             JMP   PREEMP
             JMP   WAIT
                .
                .
                .
             JMP DETACH

    KER 2:   LHLD   STACK      ;  RETURN TO EXIT ROUTINE
             SPHL              ;
             RET               ;
```

*Figure 5.13: Main Body of Kernel: I8080 Code.*

### 5.2.2   Primitives and Procedures

To support the process state diagram the following primitives are defined (these were shown in Figure 5.7):

for process multiplexing
### PREEMPT
Used by the clock ISR to stop a running process, move it to the ready to run queue (RTR) and initiate a new process

for semaphores
### WAIT (SEMAPHORE INDEX)
If the associated counter is greater than zero it is decremented by one. Otherwise the calling process is moved to the semaphore queue and a waiting process activated.

### SIGNAL (SEMAPHORE INDEX)
If the semaphore queue is empty then the counter is incremented. Otherwise a queued process is moved to the RTR queue.

for blocking and unblocking processes (by monitors)
### BLOCK AND SIGNAL (SEMAPHORE INDEX)
This primitive performs two functions — the calling process is moved to the blocked state and a SIGNAL primitive is performed on the named semaphore.

### UNBLOCK (Process Index)
This primitive moves the process from the blocked to the RTR state.

for buffer passing
### ATTACH (BUFFER QUEUE INDEX, ADDRESS OF BUFFER)
The named buffer is attached to the bottom of the queue.

### DETACH (BUFFER QUEUE INDEX, ADDRESS OF BUFFER)
The address of the buffer at the top of the queue is returned. If the queue is empty a zero is returned.

In addition to these working primitives, two initialization primitives are necessary.

### BEGIN
This primitive is called by the start-up software. It does the following:

1. Initializes the kernel data structures

2. Sets semaphore counters to zero

3. Sets semaphore buffer queues to empty

> DECLARE PROCESS (PROCESS INDEX, MICROPROCESSOR
> INDEX, PRIORITY, STACK POINTER, STARTING ADDRESS)
> This primitive initializes the process' PCB and places it in the
> appropriate RTR queue.

Initialization and start-up are considered further in Section 5.2.3.

Internally the execution of kernel primitives is facilitated by several subroutines for manipulating queues and linked lists.

Linked lists can be manipulated by two routines: ADD and REMOVE.

> ADD (QUEUE: Q, ELEMENT: PCB or Buffer Element)
> This routine attaches an element to the bottom of the queue. The address
> of the head of the queue is passed in register pair HL, while the address of
> the element is passed in register pair DE.

> REMOVE (QUEUE: Q)
> This routine removes the top element from the queue. The address of the
> head of the queue is passed in register pair HL and the element address is
> returned in register pair DE. The routine assumes the queue is not empty,
> calling routines must first test for queue empty.

Scheduling is accomplished by two subroutines: RUN and NEXT.

> RUN (ELEMENT: PCB)
> This routine places the element in the correct RTR Q. The address is
> passed in register pair HL.

> NEXT
> This routine chooses the next process to run from the appropriate RTR Q.

## 5.2.3   Initialization

The main system start-up software can be located in the private memory of the processor designated to initialize the system or in public memory. The system, on Reset, always starts with the TAS flag set; thus ISRs cannot enter the kernel. The system start-up software then calls a special system entry point to the kernel which does not test the TAS flag.

This initialization software first calls BEGIN which initializes the kernel data structures, queues and semaphores.

Next each process is declared by calling DECLARE PROCESS Finally the TAS flag is reset and interrupts are enabled to allow regular operation to begin.

Each processor must also have its own private startup software to initialize private variables and devices before enabling interrupts to start the action.

### 5.2.4  Interrupt Entry to the Kernel

Interrupt servicing was discussed generally in Section 5.1.5. The optimum solution always appears to abolish them. As the next approach, the ISRs are separated from the kernel and treated as a special type of process with restrictions.

Since the ISR procedures are required to be short and fast, they are often coded in assembly language. Since also the execution is interrupt protected, a special entry to the kernel is appropriate in which interrupts are not disabled and in which the processor state is not saved (which implies a thorough knowledge of the register states by the programmer).

If the ISRs are written in a high level language, a compatible interface must be designed. This is similar to a regular entry except that the periodic testing for interrupts is excluded.

### 5.2.5  Performance Summary

The kernel as described was found to execute primitives in the times shown in Figure 5.14 on a I8080A.

As a practical example a two processor (I808A) tightly coupled system which implemented the protocols required for packet switching (X.25) behaved as follows. Twenty kernel calls are required to send or receive a single packet. These calls coordinate the normal interactions of two ISRs and four transport processes via three monitors. At 9600 Baud the packet time is approximately 250 m secs. Kernel calls for packet transport occur on the average every 12.5 m secs. Kernel execution time is approximately 1.25 m sec. Therefore, about 10% overhead is devoted to communications over one X.25 channel. However, at a minimum, other processors are also accessing the kernel to perform time-slicing preemptions. If 3 processors are involved, and each accesses the kernel every 64 m secs for this purpose, then 10 or so kernel calls are added during each packet interval.

Thus at about 1.25 m sec per kernel call a three processor system running X.25 at 9.6 Kilo Baud has a kernel overhead of 15% (with 10% added for each additional channel)

These numbers are optimistic, e.g.,

1. Non kernel processing by other processors causes bus congestion which will slow down the processor executing the kernel — particularly if it has a low priority on the bus.

2. High priority events may cause kernel interrupts.

There always appears to be motivation for considering kernel enhancements by whatever means. Chapter 9 considers a number of ways of improving kernel efficiency.

Note that the design of a complete packet switching system which is similar in structure to the one for which these figures are quoted is described in Chapter 7.

| Primitive | # of Instructions Executed | Time (millisecs) |
|-----------|:--:|:--:|
| PRE EMPT | 180 | 0.90 |
| WAIT | 180 | 0.90 |
| SIGNAL | 130 | 0.65 |
| BLOCK AND SIGNAL | 210 | 1.05 |
| UNBLOCK | 110 | 0.55 |
| ATTACH | 80 | 0.40 |
| DETACH | 100 | 0.50 |

*Figure 5.14: Kernel Execution Times*

## 5.3  Summary

The kernel forms the central and most crucial part of the system from the performance point of view. Clearly at this time two points of view are to be maintained. First, the kernel must provide a set of primitives which permit access to common resources by processes and effective scheduling of processes by monitors. That is, the primitives must be carefully chosen to facilitate systems implementation. Secondly, the kernel must provide these mechanisms in a manner which meets certain performance criteria.

The constraint most often imposed on the kernel is its execution time (memory volume is becoming cheaper). Thus in addition to expedient code for the procedures it is important for the access mechanisms and the data structure to be designed with this response time as a driving constraint. It is often necessary to retrace the system design decisions in order to create in the structure of the system more realizable specifications for the kernel. As proposed in Chapter 6 and 7, the kernel is constrained logically from above by concurrency requirements and physically from below by hardware realities.

These two points of view are not always directly compatible. In general, the kernel design is driven by the requirements for the realization of primitives. Often the architecture of the hardware is fixed, and thus considerable ingenuity is required to meet performance criteria.

The monolithic synchronizing kernel discussed in Section 5.2 is perhaps the simplest to understand, design and implement. However, it is a bottleneck and it can monopolize the common bus.

Despite the attractiveness of monolithic kernels, newer high performance systems will demand extending the control in a variety of ways. Very little pragmatic experience is available to guide the implementation of distributed or partitioned kernels. However, little doubt exists of their importance and future systems will employ some of these concepts. Physically distributed systems in particular require extensive kernel synchronizing techniques for effective implementation. A view of such kernels will be presented in Part C, Chapter 9.

## 5.4   References

Issues surrounding kernel design and use have been covered in a number of papers by members of the Laboratory:

W. BROWN, R.J.A. BUHR, "A Concurrent High Level Language Approach to the Design of Multiple Microprocessor Systems", Proceedings of the Third Annual Symposium on Mini and Microcomputers and their Applications, MIMI 77, Zurich, June 1977, ACTA Press.

W.T. BROWN, R.J.A. BUHR, J.L. PAQUET, "Kernel Application Techniques for Multiple Microprocessor Systems", MIMI 77, Montreal, November 77, ACTA Press.

R.J.A. BUHR, W.T. BROWN, J.L. PAQUET, "PL/M as a Concurrent High Level Language", MIMI 77, Montreal, November 77, ACTA Press.

R.J.A. BUHR, J.K. CAVERS, M.H. HUI, "A Small Multiprogramming Kernel for PDP-11 Computers", 11th Canadian DECUS Symposium, Ottawa, Feb. 78.

R.J.A. BUHR, M. VIGDER, "A Firmware Kernel for a Multimicro- processor Operating System", Trends and Applications 78, Distrbuted Processing, National Bureau of Standards, Gaitersburg, Maryland, May 78.

# DESIGN AND IMPLEMENTATION OF MULTIPLE MICROPROCESSOR SYSTEMS

In this part the major emphasis will be on an extensive example to integrate the conceptual components of a concurrent multiple processor system.

Chapter 6 serves to introduce a design philosophy which accommodates the realities of practical target components and architectures. In this chapter also a summary of design principles based on this philosophy is proposed. Chapter 7 is a detailed presentation of a multiple processor implementation for a packet switching protocol. It is here that the trade-offs and accommodations can be illustrated in a working system. Chapter 8 discusses high level languages and their adaptation for program implementation in this environment. The theme throughout is on the pragmatics of the design algorithm from requirements to a working system in the context of a topical example.

Part C will provide an overview and summary of outstanding issues evoked in the preceding chapters.

Chapter **6**

# How Are Concurrent Systems Designed?

## 6.0  Introduction

In Chapter 1 an idealized top-down design philosophy was presented as a framework for thinking about system design. In Chapter 7 a detailed, practical example of the actual system design process will be presented. It is the purpose of the present chapter to build a conceptual bridge between the ideal and the practical approaches of these two chapters. In particular, this chapter defines a conceptual design process which may best be described as "edge-in", rather than "top-down" or "bottom-up". In terms of this approach, certain design principles are required to guide choices. These are summarized here and their application will be shown in the example of Chapter 7.

## 6.1  The Conceptual Design Process

Figure 6.1 provides a model of the conceptual design process. The boxes represent activities to be performed at various levels of design (requirements to be defined or design specifications to be prepared). The arrows indicate interactivity influences. This figure is a version of that provided in Chapter 1, Figure 1.3.

Starting at the top, two major activities influence all the rest. These are the functional requirements definition (Level 1) and the non-functional constraints and requirements definition (Level 2). Level 1 covers functional requirements in areas such as communications, data bases, and applications processing. Level 2 covers such aspects as flexibility, efficiency, modularity, reliability, compatibility with other products, use of standard components and techniques, etc. Level 1 influences Level 2 because of the possible need to accommodate foreseeable but incompletely defined functions. Lower levels develop the consequences of the requirements defined in Levels 1 and 2.

Level 3 (Software Architecture) is concerned with defining the software functions of the system and with partitioning these functions into active and passive modules with appropriate interfaces. Products of this activity are data flow diagrams, access

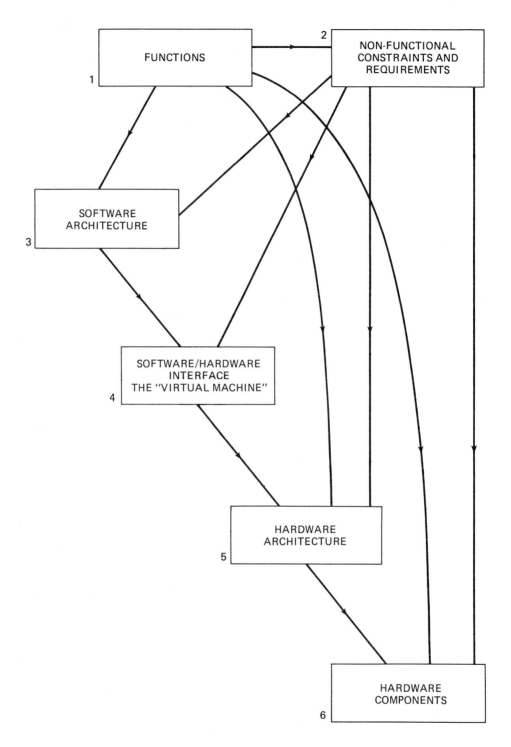

*Figure 6.1: A Model of the Conceptual Design Process*

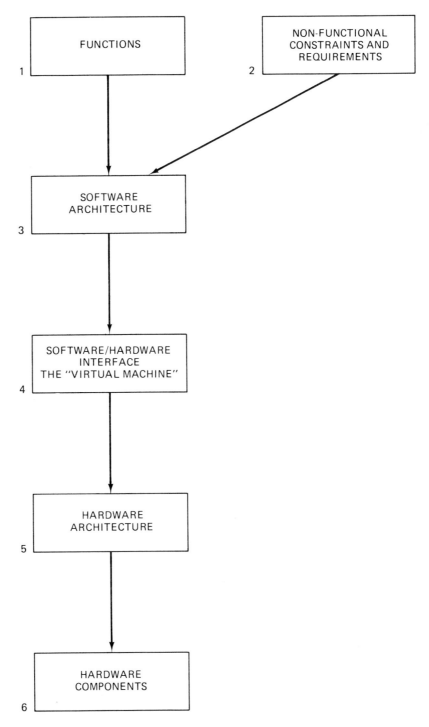

*Figure 6.2: The Top-Down Model*

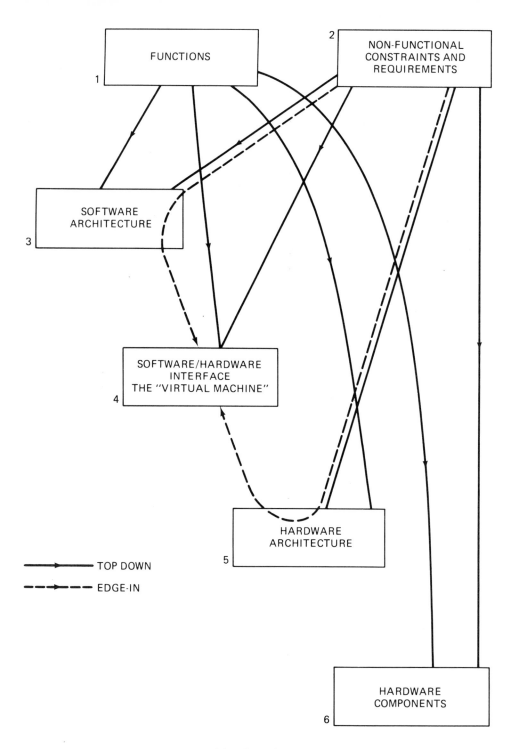

*Figure 6.3: The Edge-In Model*

graphs, module interface specifications, pseudo code for the active modules, and overall descriptions of system and module functions and operation. Level 3 directly influences only Level 4 because of the requirements imposed by particular methods of inter-process and process-to-device communications envisaged in Level 3.

Level 4 (Software/Hardware Interface) is concerned with defining the "virtual machine" which will provide the low-level mechanisms for support of the software architecture developed in Level 3; this includes the "kernel" and interrupt service routine (ISR) levels of the system. Level 4 directly influences only Level 5 because this is the only software level which directly interfaces with the hardware.

Level 5 is concerned with the hardware architecture. Aspects to be defined here include bus interfaces, method of access by processors to memory and devices, and methods of interprocessor communication. Level 5 directly influences only Level 6 because the components must be selected for compatibility with the hardware architecture.

Level 6 is concerned with hardware components, particularly the selection of chips (processor, memory management, communications, etc.).

## 6.2   The Top-Down Process

A strictly top-down design approach would follow the pattern of Figure 6.2, in which each level is fully defined before proceeding to the next lower level. In this diagram, the only inter-level influences shown are those following a strict top-down sequence. However, many of the major direct influences between the conceptual design levels are outside of this top-down structure and it is therefore not appropriate to insist that all components of this structure be in place (even if only partially) before others are considered.

## 6.3   The Edge-In Process

More appropriate for conceptual design purposes is the model of Figure 6.3 which shows only the major direct influences. Software architecture (Level 3), The Virtual Machine (Level 4) and Hardware Architecture (Level 5) can all be developed relatively independently of each other because the influences between them shown in Figure 1 are quite weak. The major influences come directly from Levels 1 and 2. And some of the influences from Level 2 can be developed more appropriately from the bottom-up. Software architecture (Level 3) has a weak influence on the virtual machine (Level 4) if standard software implementation techniques are specified in Level 2. Then the virtual machine may be designed to support these standards rather than the particular software configuration of Level 3. Similarly the virtual machine (Level 4) has a weak influence on the hardware architecture (Level 5).

In fact, requirements for the virtual machine may in many cases be almost completely determined by the "edge-in" pattern shown by the dotted lines. This is because requirements dictated by Level 2 may lead to specific choices of process multiplexing and intercommunication techniques to be used by Level 3 and made possible by Level 5. These choices may largely determine what must be done in Level 4.

## 6.4  Design Principles

Certain principles will be developed in Chapter 7 for multiple microprocessor software design to accommodate the interactions between levels 3 and 5. And certain other principles, some first proposed in Chapter 2 and others to be developed in Chapter 7, serve as a guide to good software engineering practices in Level 3. All these principles are summarized below.

### 6.4.1  Design Principles of Chapter 2

The following principles were stated and illustrated in Chapter 2:

Principle 2.1:

Transparency; create and maintain a software environment in which all concurrency problems are transparent at the process level.

Principle 2.2:

Strategic Retreat; introduce concurrency problems, at the process level, reluctantly.

### 6.4.2  Principles of Chapter 7

The following principles are stated and illustrated in Chapter 7:

1. Each concurrent (or potentially concurrent) external physical activity should have at least one process to manage its internal ramifications.

2. Protocols should be encapsulated in monitors.

3. Software for a system which may contain an (a priori) unknown number of processors should be designed so that it can be partitioned appropriately among processors without reprogramming or recompilation.

4. If all process and monitor instruction fetches must be strictly local to all processors (for efficiency, or other reasons), then a **split-process** or **split-monitor** organization is required.

5. Individual processes should not wait in multiple places for autonomous events nor should they manage multiple autonomous activities.

6. Where a process cannot avoid responsibility for managing multiple autonomous activities, it should have at most one primary responsibility and use timed waits or polling to handle the others.

7. Where it can be foreseen that the software level on one side of an interface may be eventually implemented on a special-purpose, high-performance, slave device, then that level should be designed in software as a slave level.

8. Control of startup after power-up and of restart after failure should reside in the active components of the system, namely the processes.

9. (The KISS or Keep It Simple, Stupid principle). Logical clarity and modularity of a concurrent system are enhanced by keeping the functions of individual processes simple, logically distinct from those of other processes and at the same logical level within each process; and by encapsulating logically complex or lower level functions in passive modules.

# Chapter 7

# A Multiprocessor Packet Switch

## 7.1 Introduction

This chapter illustrates the logical design process by developing in some detail the design of a multimicroprocessor packet switch. Beginning with a general statement of requirements, the chapter proceeds by a process of successive refinements through all design phases culminating in a detailed definition of all major software and hardware modules required for a practical system. Although certain simplifying assumptions are made for pedagogical purposes (for example, a rather simple link control protocol is used), the resulting system has the basic structure necessary to handle more complex requirements. Thus the reader can proceed from the high level design to particular details of implementation in the context of a single practical example.

This chapter illustrates the "edge-in" design model of Chapter 6 by showing how software architecture may be influenced by hardware architecture in a multiprocessing system.

This chapter serves not only to illustrate multimicroprocessor system design but also to introduce concepts important in distributed computing, where processor systems only communicate with each other via communications links. In particular it serves as an introduction to the subject of protocols. The extensions necessary to allow the proposed design to handle layered protocols are discussed, with particular reference to the international standard X.25 protocol.

Chapter 7 is divided into three parts. Part A defines the requirements for the example system and provides tutorial material on protocols. Much of it may be skipped by the reader with background in packet switching and protocols. Part B is central to the chapter and, indeed, to the whole book; here the logical design techniques are developed and illustrated. Part C completes the "edge-in" design process by showing how the virtual machine is designed and implemented.

Part A defines the requirements for the example. Functional requirements are developed in Sections 7.2 and 7.3; these sections also provide an introduction to packet switching and data link control protocols for those readers unfamiliar with these areas. Readers familiar with packet switching and protocols need to read only Sections 7.2.1 and 7.3.3, which summarize the material needed for Part B.

Then Part B proceeds with CHLL design. An overall software architecture for link operation is developed in Section 7.4 using data flow diagrams and access graphs. Several alternative architectures are explored, involving differing numbers of processes, and influences of different possible multiprocessor hardware configurations. The principles enunciated in Chapter 6 are here motivated and illustrated. In Section 7.5 this architecture is developed algorithmically. Section 7.6 develops the data link control module in some detail beginning with an external specification developed in Section 7.5 and proceeding through detailed internal design to actual code examples.

The actual transmission and reception of packets by interrupt service routines is discussed in Section 7.7.

Part C proceeds with the specification of the virtual machine. A hardware architecture is described in Section 7.8. A kernel to interface it to the higher levels is described in Section 7.9. Then practical details of interfacing actual software to the kernel are discussed in Section 7.10.

Finally, Part D discusses briefly (in Section 7.11) the extension required to the example system to handle practical protocols such as X.25.

## A:   REQUIREMENTS DEFINITION

## 7.2   General Functional Requirements

### 7.2.1   Introduction

This section is concerned with Level 1 of the design process described in Chapter 6.

In the interests of brevity and clarity, this section does not encompass the full range of functional requirements definition. It concentrates only on identifying the packet switching functions of a multimicroprocessor system and on describing the protocols to be used to implement them. User and application interfaces and non-functional constraints (Level 2 of Chapter 6) are treated cursorily if at all.

We are here concerned with designing the software and hard- ware architecture for the nodes of a packet switching network. Each node provides packet switching functions to co-resident node applications in a homogeneous data network of the "datagram type". In a datagram network each data packet is sent like a telegram. The network addresses of both the source and destination are sent in addition to the data (Figure 7.1)

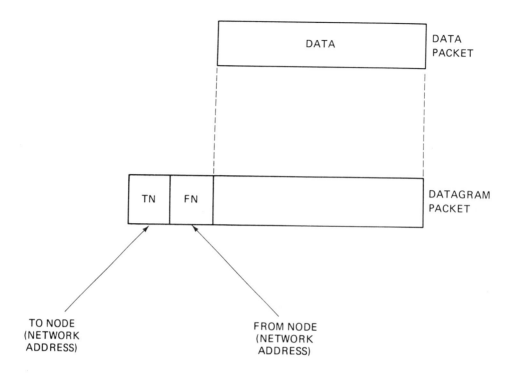

*Figure 7.1: A Datagram*

Figure 7.2 illustrates a datagram network. A packet sent by an application process in Node 1 to a corresponding application process in Node 4 is sent first to one of the nodes directly connected to Node 1, say Node 3, as dictated by the contents of an internal routing table. Node 3 receives it, examines its destination address, looks up its own routing table and sends the packet directly to Node 4. Routing tables are here assumed to be fixed at system initialization, although dynamic updating is desirable for system reconfiguration. In this network, there is no master/slave relationship between nodes; links are, in this sense, "symmetric".

Multiple microprocessors are specified to provide modularity and expandibility, with initial low cost. The key architectural problems are associated with the implementation of the protocol functions of the network.

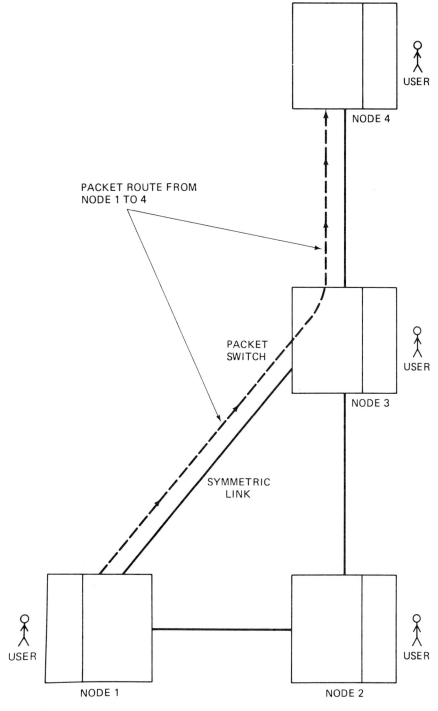

*Figure 7.2: A Datagram Packet Network Using Standard Nodes*

### 7.2.2    Protocols

Protocols control information transfer between computer systems. They are made up of sequences of messages and responses with specific formats and meanings. Protocols are required to prevent misunderstandings between computer systems due to transmission errors, to autonomous computer system startups and shutdowns and to data overload conditions. They are implemented by computer programs which are required to be functioning correctly. A malfunctioning protocol program may lead to uncorrectable misunderstandings between computer systems at the level of that protocol; higher level protocols are then required which may themselves malfunction, and so on. At some level, reliability must be assumed or guaranteed.

For our datagram network, transmission errors will be detected by adding a redundant check code to each datagram packet at the sending end and recomputing it at the receiving end; failures will result in packet rejection by the sender. An appropriate and popular check code which has been adopted by International Standard is the 16 bit CRC (cyclic redundancy check) which can be computed and added to the data stream by software or by a communications control chip. Thus transmission errors are equivalent to packet losses.

During system startups and shutdowns, during off-the-air periods, and during data overload conditions (all buffers full) a system will be deaf to reception. All these conditions will thus appear to the sender to have the same effect as transmission errors, namely packet loss.

The sole objective of the protocol to be described in the next section is to recover from all packet losses. Efficiency considerations are ignored, in order that the main principles of implementing protocols on a multimicroprocessor system can be explored without burying the discussion in complicated detail. Extensions for practical protocols are discussed later (Section 7.11).

In our datagram network, the message content of the protocol itself will be imbedded in a header and the check code imbedded in a trailer attached to every datagram packet, as shown by Figure 7.3. The datagram packet is thus transmitted as a data link "frame". As will be shown, there is a requirement for non-data frames for link control purposes; in such frames. the length of the data portion is zero. The link protocol will now be discussed in the next section under two headings:

1.  Link Initialization

2.  Link Operation

### 7.3    Data Link Control Protocol

This section not only specifies the protocol requirements for our datagram network, but also serves as an introduction to protocols. Because it illustrates so many of the problems of protocols, we begin with a discussion of link initialization.

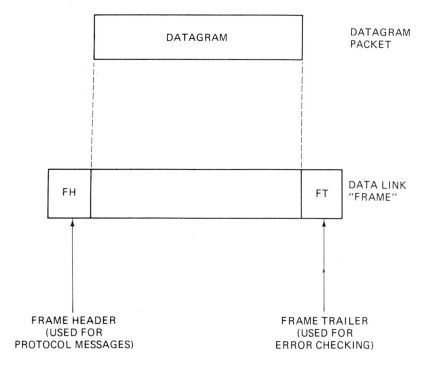

*Figure 7.3: A Data Link "Frame"*

### 7.3.1  Link Initialization

For readers unfamiliar with protocols, this section provides considerable introductory material on the nature and pitfalls of protocols.

After power-up or in recovering from a "crash", before starting to send or receive datagram frames, a node at the end of a "symmetric" link must ensure that the other node is ready and that there is mutual understanding of where to start in the bi-directional packet stream. Otherwise, a node could be sending data to a failed or inactive node or to one which is active but not expecting the particular frames sent. In the latter case, which could occur if one end of the link "crashes" and then recovers without the other end being aware of it, frames could be permanently lost (this point will be explored in more detail later).

The link initialization requirements are developed below by considering and rejecting a sequence of plausible but incorrect protocols of increasing refinement until finally a correct one is achieved. In this way the reader is gradually introduced to the subject of protocols.

### 7.3.1.1 A Naive Protocol

Suppose that after the power is turned on or after a crash each node comes on the air by transmitting a special START frame to every node to which it is directly linked. This frame is identified by a special code in the FH field. Then a very simple (and naive) protocol is described by the following algorithm, to be executed by each node:

1. Enter LINK-DOWN state

2. Send START

3. Wait for START

4. When START received, enter LINK-UP state and begin normal operation.

This protocol has faults, but before discussing them, we turn to an alternate representation of the protocol as a finite state machine (FSM) as shown in Figure 7.4. The node cycles through three states: 0:LINK-DOWN, 1:WAIT-FOR-START, and 2:LINK-UP.

Transitions between states are triggered by events; here, events are either local requests for link status changes or the transmission or reception of frames. Transitions

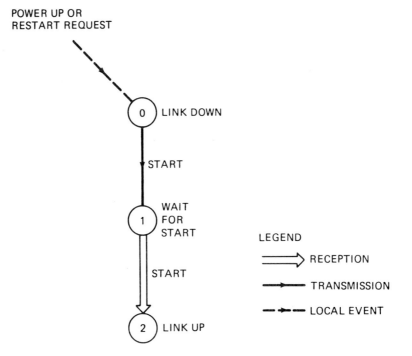

*Figure 7.4: A Naive Startup Protocol*

caused by transmissions are shown as single lines and by receptions as double lines; transitions induced by events local to the node but outside the protocol are shown by dotted lines.

This naive protocol illustrates the problem of uncertainty associated with distributed systems and noisy communications. Knowledge of the current state of a remote node is always uncertain because of message time delays and possible failures in the link to the remote node or in the node itself. Protocols must handle such uncertainties by making assumptions about the states of remote nodes based on previous messages and then being prepared to change the assumptions if the responses from the remote nodes are inappropriate. In Figure 7.4 two possibly inappropriate assumptions are made about the two nodes at the end of a link (Figure 7.5a):

1. Each will be in State 1 when the other's START frame arrives.

2. When either enters State 2, it can be sure that the other will eventually enter State 2.

However, there is clearly a possibility that a START frame could be missed because the node was not in State 1 when it arrived or that a START frame could be lost entirely because of a link failure (Figure 7.5b). Either possibility invalidates the above assumptions and there is no provision in the protocol to change the assumptions.

### 7.3.1.2   A First Refinement of the Naive
### Protocol

To cater for these possibilities, a timeout may be introduced in State 1, as shown in Figure 7.6; this amounts to a changed assumption about the state of the remote node after a period of time with no response. Such timeouts are necessary features of protocols, in general. Following the new protocol, each node repetitively sends START frames until it receives one itself, when it enters the LINK-UP state. It is no longer possible for a node to "hang" forever in State 1 because a frame was missed or lost. However, a new possibility has been introduced, that of one node being trapped in an endless timeout cycle waiting for a lost START frame from the other node (Figure 7.7).

### 7.3.1.3   Second Refinement of the Naive
### Protocol

An attempt to patch the protocol is shown in Figure 7.8, where receipt of a START frame in State 2 is recognized as an indication that the remote node did not receive the one sent earlier. A quick transition through a new state (State 3) and the sending of another START frame fixes that problem (Figure 7.9). Now everything is fine as long as it is never possible for one node to be in State 2 while the other is in State 3. But Figure 7.10 shows that this state combination is possible and that a cyclic deadlock results.

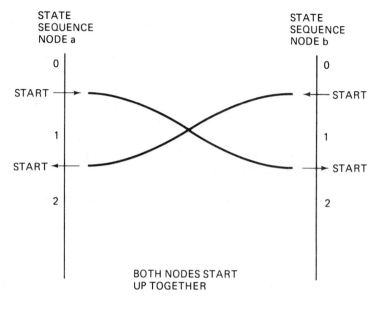

(a)  IDEAL

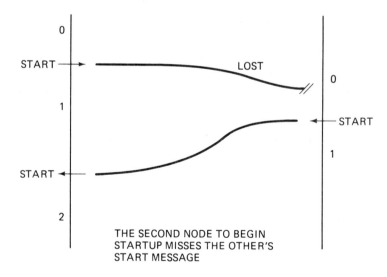

(b)  FAILURE OF THE NAIVE PROTOCOL

*Figure 7.5: Operation of the Naive Startup Protocol*

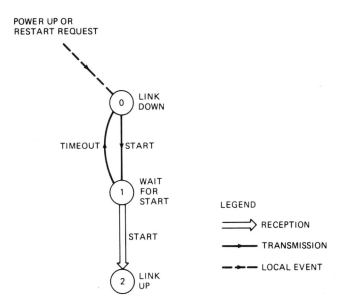

*Figure 7.6: A First Refinement of the Naive Startup Protocol*

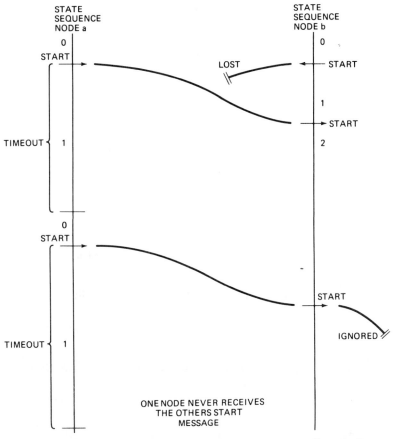

*Figure 7.7: Failure of the First Refinement of the Naive Startup Protocol*

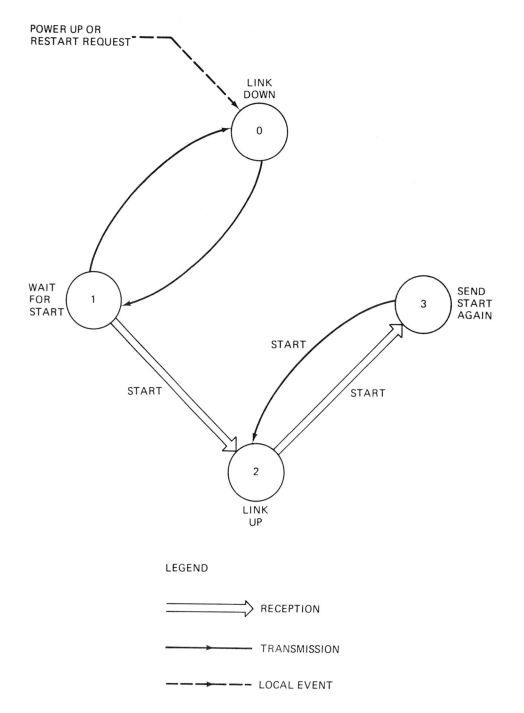

*Figure 7.8: A Second Refinement of the Naive Startup Protocol*

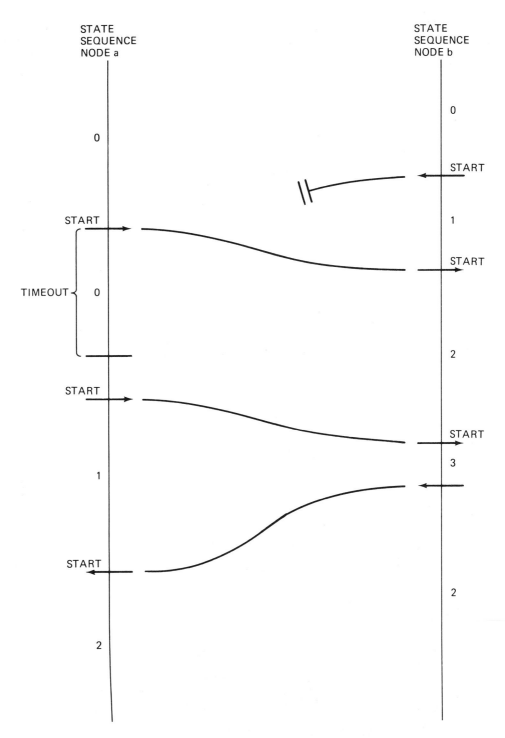

*Figure 7.9: Operation of the Second Refinement of the Naive Startup Protocol*

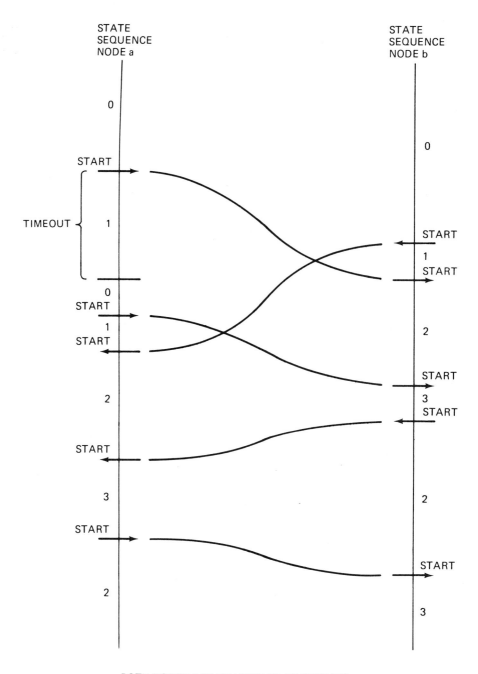

BOTH NODES ARE TRAPPED IN AN ENDLESS
LOOP BECAUSE THEY CANNOT DISTINGUISH
"REAL STARTS" FROM "ACKNOWLEDGEMENT STARTS"

*Figure 7.10: Failure of the Second Refinement of the Naive Startup Protocol*

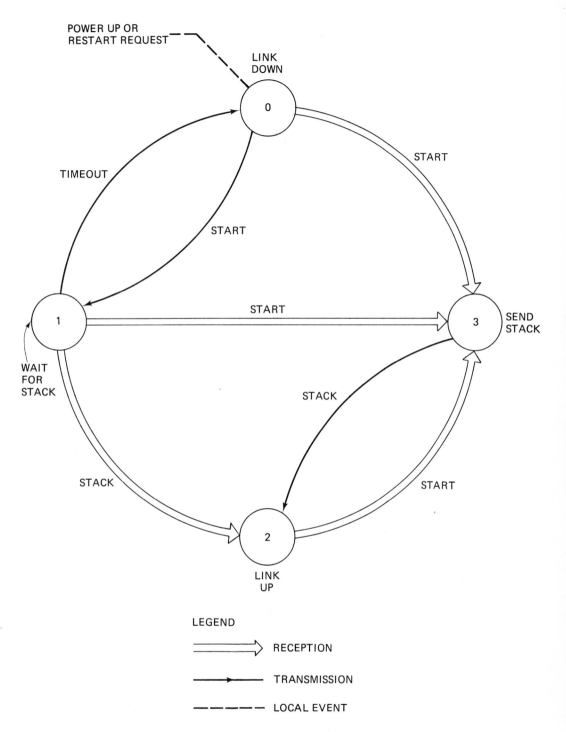

Figure 7.11: A Correct Link Startup Protocol

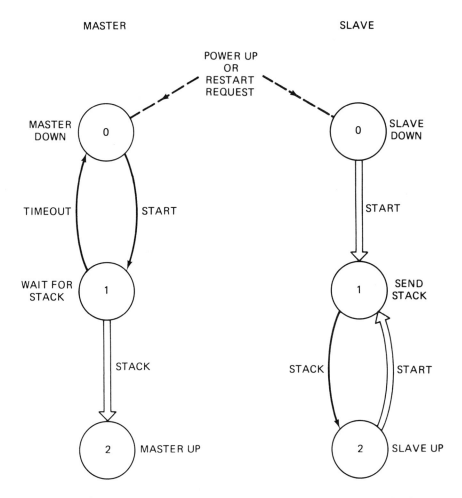

NOTES:  IF <u>BOTH</u> MASTER AND SLAVE ARE IN STATE '0', THE LINK IS DOWN
        IF <u>BOTH</u> MASTER AND SLAVE ARE IN STATE '2', THE LINK IS UP
        OTHERWISE THE LINK IS <u>COMING UP</u>

LEGEND

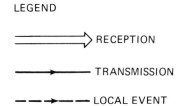

*Figure 7.12: An Alternate Correct Link Startup Protocol*

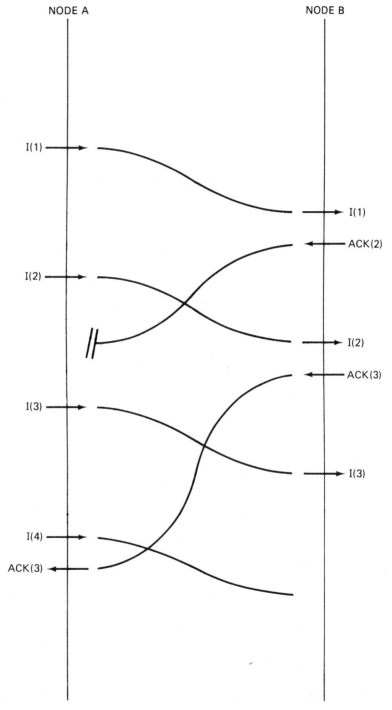

*Figure 7.13: Implicit Acknowledgement in a SPAR Protocol*

### 7.3.1.4    A Correct (PAR) Protocol

What is required is the addition of a new message to the protocol, a STACK (start-ack) message, which can be used to break the endless loop between States 2 and 3. Now the rule is (Figure 7.11) whenever a node receives a START message it responds with STACK instead of START. Now only one transition from State 2 to State 3 and back is possible. In State 1, receipt of either START or STACK is accepted as a sign of remote system activity, because we know we can recover from any failure of the remote end to receive START by the exit from State 2. For uniformity, acknowledgements are triggered by receipt of a START in any of States 0, 1 or 2. The reader should be able to convince himself of the correctness of the protocol by drawing timing diagrams showing various possibilities. Such a protocol is known as a PAR (Positive Acknowledgement Retransmission) protocol.

### 7.3.1.5    Some Protocol Requirements

This correct protocol illustrates several important requirements of protocols.

1.  Timeouts and retransmissions must be included.

2.  Unambiguous positive acknowledgement messages are required in the protocol for all autonomously arriving protocol messages.

3.  The "last message" of an exchange cannot be critical to the successful operation of the protocol. If it were, then the sender of the last message would need to wait for its acknowledgement to be sure that it was received correctly, and it would no longer be the last message. The 2-3-2 cycle in Figure 7.11 illustrates this feature: at no point in the cycle is it ever required that the "last" STACK message be explicitly acknowledged. It is implicitly acknowledged by receiving no more startup protocol messages.

4.  A standard initial state is required which is always entered in the same way from other levels of the system. Otherwise, additional interactions between states of the different levels must be taken into account in validating the protocol.

5.  Exhaustive analysis is required to ensure that no deadlocks, endlessly repetitive loops or inconsistent actions occur. More precisely the analysis must consider all possible states of the link, including the protocol states at either end and the influence of frames in transit. It is not sufficient to consider only the protocol states at the ends of the link because frames in transit can change these states (or fail to change them as expected, in the case of frame loss).

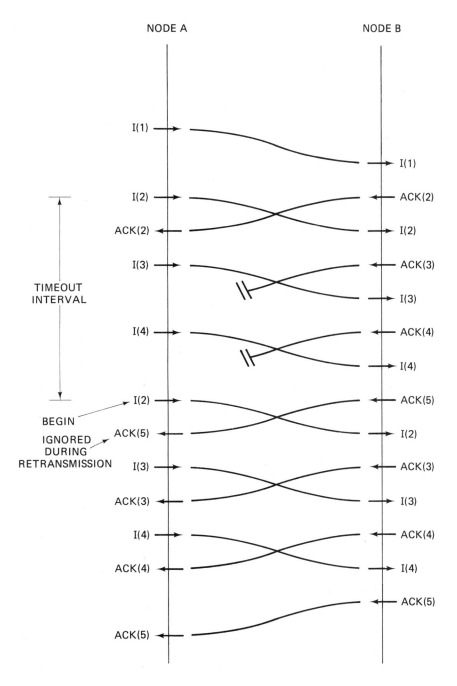

*Figure 7.14: Retransmission in Acknowledgement and Timeout in a SPAR Protocol*

##### 7.3.1.6    *Other Startup Protocols*

Another approach to symmetric startup protocols is to split them into master and slave sub-protocols as shown in Figure 7.12. The master sends commands and receives acknowledgements. The slave receives commands and sends acknowledgements. The two sub-protocols must both be in the "up" state before the link is declared "up". The overall logic of this approach is somewhat more complicated because two independent protocols must now be coordinated. However, they are adaptable to master/slave link organization and they are found in practice in the link level protocols of public packet switched networks. For example, one version of the link level of X.25 incorporates such a split startup protocol.

### 7.3.2    Link Operation

#### 7.3.2.1    *SPAR Protocols*

Links in our datagram network are required to support full- duplex transmissions. Therefore it is not permissible in general to hold up frame transmission while waiting for acknowledgement of receipt of a frame by the remote node. To support a stream of frames in both directions a so-called SPAR (Sequencing Positive Acknowledgement Retransmission) protocol is required. SPAR protocols attach sequence numbers to packets. Each packet is positively acknowledged by its sequence number. And unacknowledged packets are retransmitted.

SPAR protocols must satisfy the protocol requirements developed in Section 7.3.1.5. There, it was pointed out that unambiguous positive acknowledgements are required for each protocol message. In a SPAR protocol this is accomplished by appending a numbered protocol message to each transmitted data frame. Acknowledgement messages reference this number. If the numbers are in sequence then an implicit but unambiguous positive acknowledgement of a sequence of messages may be accomplished by referring only to the last number of the sequence in the acknowledgement message. SPAR protocols also employ timeouts to recover from frame losses.

In a SPAR protocol a frame stream is transmitted without waiting for individual frame acknowledgement. Frames in the stream are uniquely identified by numbering them sequentially from zero modulo some maximum number K. Positive acknowledgement of transmitted frames takes the form of the number of the next frame expected, modulo K (implicitly acknowledging receipt of all frames with lower numbers, modulo K). Acknowledgements are attached by the receiver to data frames or sent separately in explicit ACK frames, which are simply empty data frames, if there is no data to be sent. A copy of each unacked frame is kept by the sender in a timeout queue until it is acked. There may be at most K-1 unacked frames in this queue after which no new frames may be transmitted until further acknowledgements are received; otherwise sequence numbers would not be unique.

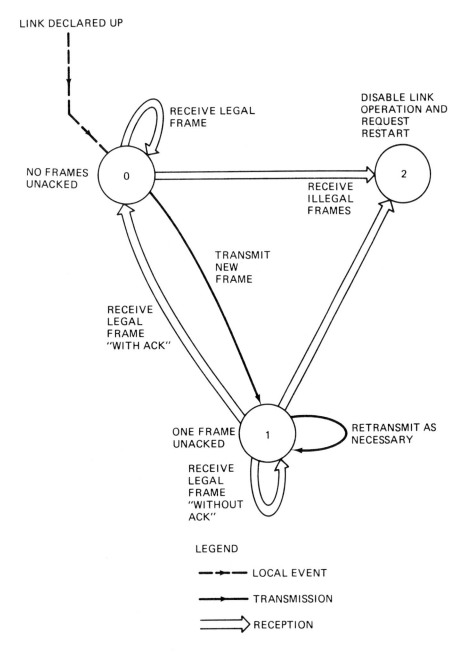

LEGEND

— ▸ — LOCAL EVENT

——▸ TRANSMISSION

⟹ RECEPTION

NOTE:
"WITH/WITHOUT ACK" — MEANS RECEIVE COUNT
DOES OR DOES NOT ACK THE CURRENTLY UNACKED FRAME

*Figure 7.15: A Simple SPAR Protocol*

Frames remaining in the timeout queue are retransmitted periodically. At the receiving end error-free frames with sequence numbers equal to the next expected sequence number (NESN) are accepted and acked (and the NESN incremented). Frames with CRC errors are simply discarded, unacked. Frames with unexpected sequence numbers or acknowledgement numbers indicate failures of some kind and will usually trigger special actions, although in certain cases they may be simply discarded (see further discussion below). Any system shutdown or system failure outside the protocol should inactivate the protocol until the condition is corrected, when link restart will be triggered.

Correctly functioning SPAR-type protocols have been shown to have the following desirable properties; they never lose frames, duplicate frames, fail to deliver frames or deliver frames out of order. Note that loss is distinguished from failure to deliver; the former implies positive acknowledgement of an undelivered frame.

However, a failed SPAR protocol (due to a computer hardware or software error, for example) may lose, duplicate or fail to deliver frames. For example, if the failure happens to change the next expected sequence number to a value below the sequence number of the oldest unacked frame, then no further frames will ever be acked and the link will be stuck in an endless retransmit/reject loop. All frames fail to be delivered in this case. If the next expected sequence number is changed to a value in the middle of the range of sequence numbers of unacked frames, then some frames will be lost (acked but never delivered) and some frames will be duplicated. And so on. Timely inactivation of the protocol when a system failure occurs and correct restart of both ends of the link is thus crucial to the success of the protocol. In general, failure of the protocol itself cannot be made invisible to higher levels.

The operation of a SPAR protocol is illustrated in Figures 7.13 and 7.14:

(a) Figure 7.13 depicts error free transmission of four frames. The acknowledgement of information frame I(1) is received with errors and discarded. The acknowledgement of frame I(2) is received error free and acknowledges frames I(1) and I(2) allowing the transmitter to reset the timer associated with frame I(1). In this protocol, if the acknowledgement is not received prior to timeout, the transmitter "goes-back-N" and retransmits all frames beginning with the frame which caused the timeout.

(b) Figure 7.14 depicts a situation in which frames I(2), I(3) and I(4) are retransmitted because both the explicit and implicit acknowledgements of frame I(2) did not arrive on time. Notice in figure 7.14 that although A(5) arrives implicitly acknowledging I(2), I(3) and I(4) it is ignored because at this time the transmitter has "forgotten" that I(2), I(3) and I(4) have already been sent once.

### 7.3.2.2  A Simple SPAR Protocol

To simplify the first pass at a system design, a very simple, degenerate version of the SPAR protocol will be implemented in which all SPAR counting functions are performed, but the number of unacked frames in each direction is restricted to one. Then frame transmission/retransmission reduces to repetitive transmission of each frame until it is acked. The resulting protocol is simple enough that in the development of a system design to implement it, the discussion can concentrate on issues rather than details. With this protocol all unexpected sequence or acknowledgement numbers trigger link restart. Again a simplification has been made; a more general SPAR protocol would simply ignore unexpected packets with sequence numbers absolutely greater than the next expected sequence number. Such packets probably imply a missed packet due to a transmission error, rather than a protocol failure, and therefore the retransmission mechanism can probably be relied upon to provide recovery. In our simple case, such packets indicate a protocol failure. And finally, our simple protocol will have no provision for flow control. That is, there is no way for the receiver to tell the transmitter explicitly that it is too busy to receive.

This simple link operation protocol is illustrated in FSM form in Figure 7.15. The protocol has two operational states: State 0, in which no frames are unacked, and State 1, in which one frame is unacked. These two states may be represented by the two values of a boolean variable "busy", because in State 0 the link is not busy and in State 1 it is busy retransmitting. The acknowledgements are by sequence number as for a general SPAR protocol.

In this protocol, retransmission is performed on some, as yet unspecified, periodic basis. The retransmission could be simply performed by an endless program loop (inefficient) or on a timeout basis triggered by a clock (better). In either case the logical concept is "send-and-wait": send a packet then wait for an ack — if no ack is received after a certain time, then send again. This operation is illustrated by Figure 7.16.

In Figure 7.16, frame I(1) and its acknowledgement are transmitted correctly. Frame I(2) is transmitted correctly but its acknowledgement is lost due to errors. Since no acknowledgement arrives before the timeout interval expires, I(2) is retransmitted. The second transmission of I(2) is lost so I(2) is transmitted a third time. In this case, transmission of I(2) and the corresponding acknowledgement are successful.

It might be thought that with this simple "SEND-AND-WAIT" concept, sequence counts are not necessary. That sequence counts are necessary (modulo 2 at least) can be seen by considering that if acknowledgements are not sequence numbered then a lost acknowledgement can result in retransmitted copies of old frames being interpreted as new frames. This situation is illustrated by Figure 7.17:

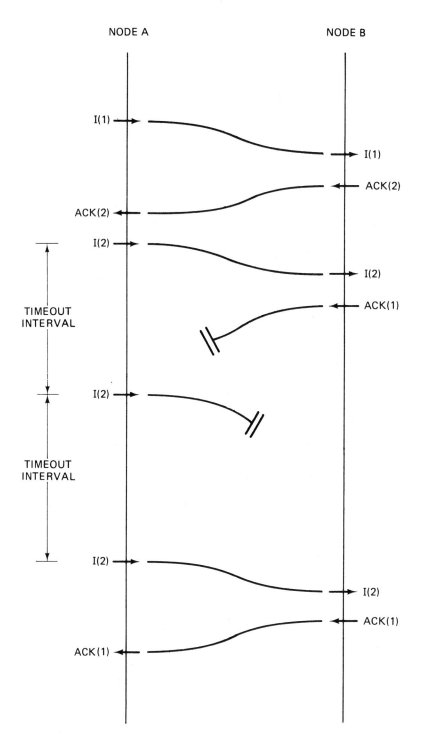

*Figure 7.16: Illustration of the Simple Link Operation Protocol*

— sender sends frame I(1)

— receiver receives frame I(1) correctly

— receiver sends acknowledgement of frame I(1); this is labelled ACK (1) because there are no other possible sequence counts

— acknowledgement is corrupted by noise and is not received correctly by the sender

— the timeout interval expires so the sender retransmits frame I(1)

— receiver receives a second copy of frame I(1) correctly which it must treat as a new and distinct frame since it is unaware that a timeout has occurred

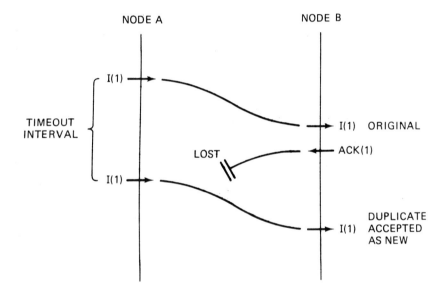

NOTE:  IN THIS EXAMPLE SEQUENCE NUMBER 1
IS THE ONLY POSSIBLE SEQUENCE NUMBER.

*Figure 7.17: Erroneous Reception of Duplicates When There is No Sequence Numbering*

### 7.3.3   Protocol Requirements

To implement the link startup and operation protocols, datagram packets must be encapsulated into data link "frames" (Figure 7.18). Two types of frames are defined: data and control frames. The frame header (FH) of data frames contains send 'and receive counts. The send count field in FH provides the sequence number of the current frame. The receive count field provides the expected value of the send count in the next incoming frame; it provides positive acknowledgement of all frames with smaller sequence numbers.

An explicit ACK frame is just a standard sequence-numbered data frame with a zero length data field. It is used to acknowledge a received frame when there is nothing to be sent. Special unnumbered control frames START and STACK are required for link startup. These frames bear no data or sequence counts. Therefore the FH requires a "frame-type" field to distinguish these control frames from data frames. The frame trailer field of all frames contains the redundancy check code.

Figure 7.19 depicts the combined startup and operation protocol. States SO, S1, S2 and S3 are the startup states of Figure 7.11. States RO, R1 and R2 are the operational states of Figure 7.15. Care must be taken in linking different protocol layers to ensure that new interactions between states of the different layers are not introduced. Otherwise the protocol would require re-validation in the more complex new environment. In this case, new interactions are avoided by ensuring that on all transitions into state RO from the startup layer, all operational protocol variables are re-initialized to ensure the system is starting afresh.

The combined protocol of Figure 7.19 will be implemented for each link. The question of inter-link influences now arises. The internal mechanisms of each node will ensure that datagrams are forwarded over the correct next link, if it is operational. However, if the next link is down, datagrams cannot be forwarded. And yet their receipt by the node will already have been acknowledged by the link operation protocol of the receiving link. A higher level protocol must clearly come into play which will guarantee end-to-end transmission of datagrams. Consideration of this end-to-end protocol is outside the scope of our example. Without this protocol, datagrams will be lost on an end-to-end basis if any of the links fail.

Implementing Figure 7.19 requires two protocol state variables which may take on the following values:

```
OP   =   RO  :  IDLE
         R1  :  BUSY
         R2  :  OFF

STRT =   SO  :  DOWN
         S1  :  WAIT
         S2  :  SEND
         S3  :  UP
```

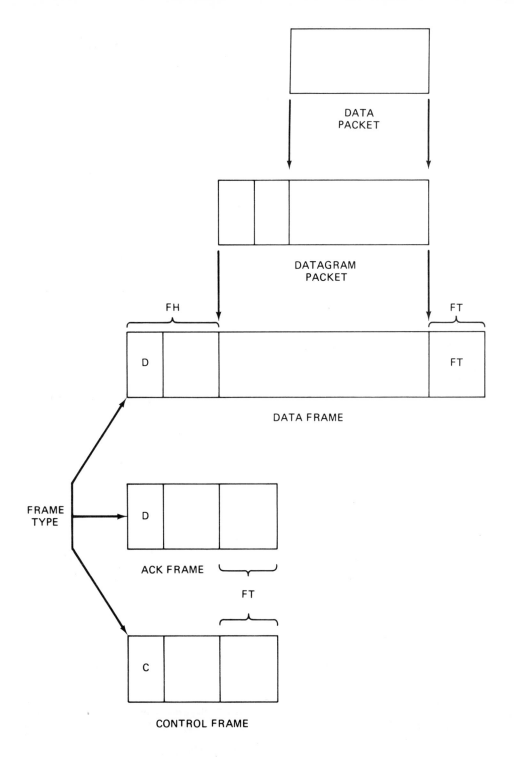

*Figure 7.18: Frame Types*

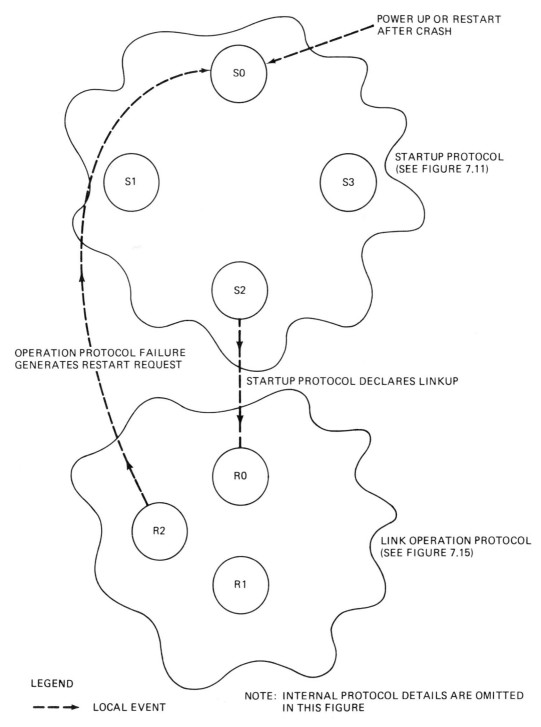

POWER UP OR RESTART
AFTER CRASH

STARTUP PROTOCOL
(SEE FIGURE 7.11)

OPERATION PROTOCOL FAILURE
GENERATES RESTART REQUEST

STARTUP PROTOCOL DECLARES LINKUP

LINK OPERATION PROTOCOL
(SEE FIGURE 7.15)

LEGEND

- - - →   LOCAL EVENT

NOTE: INTERNAL PROTOCOL DETAILS ARE OMITTED
       IN THIS FIGURE

*Figure 7.19: The Combined Protocol*

The transition rules for these states are specified by Figure 7.19. Which protocol is active may normally be inferred from the state variables themselves. If STRT is anything but UP then STRT is active and OP must be OFF. If OP is anything but OFF, then OP is active and START must be UP. However, if OP is OFF and START is UP then the state is apparently ambiguous : the link could either be coming up (in which case the startup protocol is active) or going down (in which case operation protocol is active). The apparent ambiguity is easily resolved by ensuring that STRT = UP is always indivisibly and automatically followed by OP = IDLE when the link is coming up.

## B:   CHLL DESIGN

## 7.4   System Architecture

This section is concerned with Level 3 of the design process described in Chapter 6 and with the interactions between Level 3 and Level 5.

In this section, an overall system architecture is developed in terms of data flow diagrams and access graphs. The development process is one of successive refinements of "pictures" of module interactions as expressed by access graphs. The process may be described as "edge-in" (rather than "top-down" or "bottom-up") development, because we proceed "inwards" from the external interfaces (the system "edges"). Alternative architectures are considered before a final one is selected, bearing in mind the possible multiprocessor nature of the final system.

### 7.4.1   Functions and Data Flow

An important first step in developing a system architecture is the definition of the functions and of the data flow patterns between functions.

The basic functions of the system from the point of view of link startup and operation may be enumerated as follows:

1. Manage the startup and operation protocols for each link as a function of local system requests and of frame transmission and receptions over the link.

2. Provide datagram reception and transmission services to local applications.

3. Provide datagram transfer facilities between links.

These functions are shown in a basic data flow diagram in Figure 7.20.

On the reception side, character or bit streams arriving from the physical link are assembled into frames which, after they have been checked for validity, are stripped of

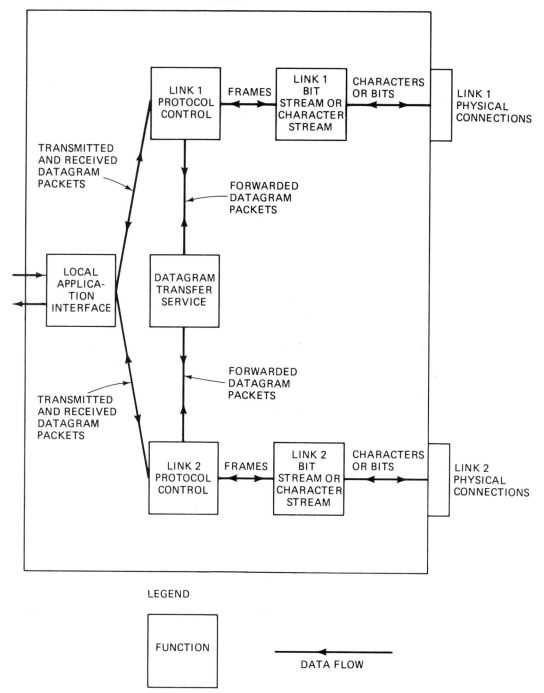

NODE

LINK 1
PROTOCOL
CONTROL

FRAMES

LINK 1
BIT
STREAM OR
CHARACTER
STREAM

CHARACTERS
OR BITS

LINK 1
PHYSICAL
CONNECTIONS

TRANSMITTED
AND RECEIVED
DATAGRAM
PACKETS

FORWARDED
DATAGRAM
PACKETS

LOCAL
APPLICA-
TION
INTERFACE

DATAGRAM
TRANSFER
SERVICE

FORWARDED
DATAGRAM
PACKETS

TRANSMITTED
AND RECEIVED
DATAGRAM
PACKETS

LINK 2
PROTOCOL
CONTROL

FRAMES

LINK 2
BIT
STREAM OR
CHARACTER
STREAM

CHARACTERS
OR BITS

LINK 2
PHYSICAL
CONNECTIONS

LEGEND

FUNCTION

DATA FLOW

*Figure 7.20: Data Flow Diagram for a Node with Two Links*

**174**

the link protocol fields and passed on as packets to either local applications or to other links for onward transmission. On the transmission side, packets are passed directly to the appropriate link transmission/reception functions for assembly into frames for subsequent transmission.

## 7.4.2 Finding the First Processes

The arrows in this diagram represent data flow which will be managed somehow by one or more processes. Our task now is to define an appropriate set of processes and a supporting system architecture to manage this data flow in a logically clean and efficient manner. How many processes are required? What should they do? How are they to coordinate their activities? What data buffering mechanisms are required? All of these questions and more must be answered.

A first guiding principle for design which provides a start on finding processes is the following:

Principle 1

Each concurrent (or potentially concurrent) external physical activity should have at least one process to manage its internal ramifications.

This principle leads not only to clean, logical separation of the management of external activities but also to maximum flexibility and to potential efficiency in an environment which may contain one or many processors.

In this example, the basic external physical activities are the transmission and reception of frames over the two directions of each full duplex link. Therefore the above guiding principle requires at least two processes in the system for each full duplex link (one per direction). We defer for the moment the question of what these processes will do.

There will also be application processes as appropriate for the application areas and associated operator interfaces. Although these are outside the direct scope of this section, we may pause to note that the above guiding principle requires at least one and possibly more than one process for each active operator interface station. One process is required if operator interaction is conducted by alternating dialogue as is usually the case with computer terminals. More processes may be required for more complex forms of interaction.

As a matter of interest, note that this guideline would lead to one process per caller in a computer-based telephone switching system and to a potentially enormous number of active processes. This is, in fact, a perfectly logical organization which has been implemented in some practical telephone switching systems. There are obvious efficiency drawbacks, but alternate organizations which assign steps in call processing to different processes, which handle these steps for all callers, create their own problems. Such alternate schemes fall under the heading of "software pipelining" and the problems they create are primarily associated with error recovery (because handling of a single call is no longer centralized in one process). We postpone further discussion of the interesting subject of software pipelining until Section 7.4.9.

Are any additional communication processes required than the two per physical link suggested above? To answer this question, consider the frame data flow over a full duplex link depicted by Figure 7.20. The link operation protocol specified in the previous section may generate a many-to-one frame-to-packet relationship in either direction. It seems logical, at least as a first pass, to assign one process to handle each direction of frame activity and to introduce no new processes at this stage.

### 7.4.3    Edge-In Development of an Access Graph

We have thus arrived at the rather basic access graph shown in Figure 7.21(a) by an approach which may be described as "edge-in", if it is considered that the external autonomous activities are at the system "edges". Note the requirement to provide a mechanism for requesting link status changes. We defer for the moment the question of who makes these requests. We shall now develop the internal structure of the "amorphous conceptual glob" representing the undefined interface in this figure by continuing this "edge-in" development. We shall concentrate first on normal link operations.

What is the nature of the interface required between the processes in Figure 7.21(a)? First, the interface must provide each application process and each frame transmission process (labelled OUT in the figure) with a place to pick up new packets or frames. This may be done on either a wait or poll basis. Waiting is fine if the process has nothing to do otherwise, as is the case with all the frame-OUT processes. However, application processes should not in general have to wait for autonomous input. Accordingly the application process interface should provide a no-wait packet pick-up call which can be polled periodically by application processes. In either case packet pick-ups, deposits and transfers should be mutually exclusive activities. The appropriate structure is a monitor, which provides protected calls which can include waits if required.

An additional reason for using monitors is the very natural and logical way in which they can be used to encapsulate the event-driven FSMs which control the link protocols. The reader will recall that state changes in these FSMs (Figures 7.11, 7.15 and 7.19) are triggered by transmissions, receptions and local requests. If the FSMs are encapsulated in a monitor, then the monitor calls themselves provide the events and the strictly controlled sequential access to the internal variables of the monitor by the monitor procedures guarantees that events will not affect the FSMs until all state variables are consistent.

Based on the above arguments, we may now state another guiding principle for design:

Principle 2

Protocols should be encapsulated in monitors.

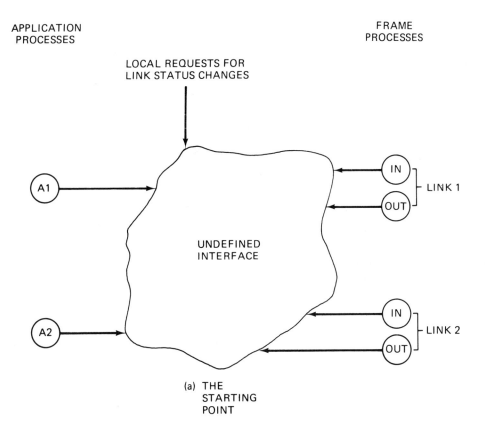

APPLICATION
PROCESSES

FRAME
PROCESSES

LOCAL REQUESTS FOR
LINK STATUS CHANGES

A1

IN

OUT

LINK 1

UNDEFINED
INTERFACE

A2

IN

OUT

LINK 2

(a) THE
STARTING
POINT

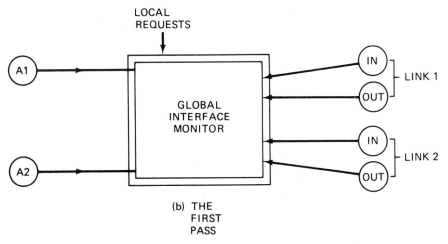

LOCAL
REQUESTS

A1

IN

OUT

LINK 1

GLOBAL
INTERFACE
MONITOR

A2

IN

OUT

LINK 2

(b) THE
FIRST
PASS

*Figure 7.21: Edge in Development of an Access Graph*

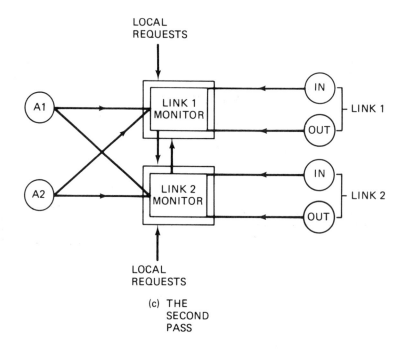

(c) THE
SECOND
PASS

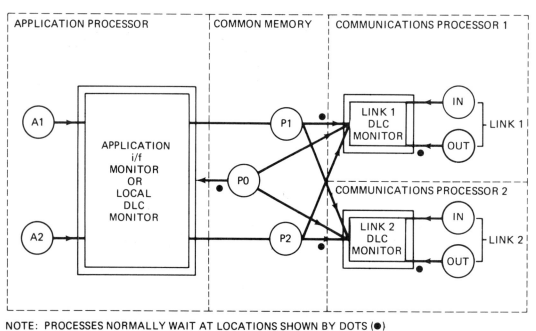

NOTE: PROCESSES NORMALLY WAIT AT LOCATIONS SHOWN BY DOTS (●)

(d) THE
FINAL
CHOICE

*Figure 7.21: Edge in Development of an Access Graph (Cont'd)*

178

Figure 7.21(b) shows a possible architecture. A single global interface monitor encapsulates all packet and frame pickups, deposits and transfers. The monitor performs all protocol functions for all links as well as inter-link frame packet forwarding. However, this structure is excessively monolithic. There is no good reason for combining all links into one monitor except to share code. And in a multiprocessing environment where each link may possibly be handled by a separate processor with its own program memory, this sharing may not even be possible or desirable. For bus efficiency reasons it may be desirable for the low level code for each link reside in the private memory of separate processors. Ultimately, the entire interface for each link may be on a single, special purpose chip.

Based on the above discussion we may now state the following third design principle.

### Principle 3

Software for a system which may contain an (a priori) unknown number of processors should be designed so that it can be partitioned appropriately among processors without reprogramming or recompilation

Applying this principle to Figure 7.21(b), a partitioning of the single monitor is clearly indicated. A first attempt is shown in Figure 7.21(c). Here there is a monitor for each link and inter-link frame forwarding is accomplished by inter-monitor calls. Does this structure accomplish the desired partitioning according to Principle 3? Certainly the link monitors can be separately compiled and then, if required, located without recompilation in the separate private memories of individual processors. However, once there, they can't call each other unless external access is provided to private memory, which is not generally the case in multiprocessor configurations. Furthermore a deadlock will occur with this arrangement if the monitors call each other while both gates are locked.

To provide the necessary partitionability, and to avoid the deadlock problem we introduce an additional layer of simple packet transport processes as shown in Figure 7.21(d) which, because of their simplicity, can presumably always reside in common memory, thus remaining executable by any processor. These transport processes provide the forwarding function between links and the pick-up and delivery functions between links and applications, all on a datagram packet basis. These processes are simple, because all protocol functions are performed by the monitors.

Each of these proceses is responsible for waiting for input on its "own" link monitor, as shown in the figure. As for the monitors, there is one monitor to provide an interface to local application software and one Data Link Conrol (DLC) monitor for each physical link. These monitors may all be made to look alike externally to the transport processes, with obvious advantages for software modularity, if we view the application interface monitor as a DLC monitor for an internal "link" to local applications. Internally this "local DLC" monitor does not need to implement a link protocol; it simply passes packets on, untouched. However, this difference in internal complexity need not be visible externally. Link operations and applications can now reside on different processors, if required.

We have arrived at the basic access graph of Figure 7.21(d) by a straightforward argument, following a few simple principles. We should now review the various possibilities of Figure 7.21 from several different viewpoints. Several questions naturally arise. Are further module splittings possible or desirable? How do the various options stand up from an efficiency viewpoint? What should be the responsibilities of processes and how should they wait or poll for results? How are buffers to be managed? How are starting and restart to be controlled? We consider them and other questions in the following sections.

### 7.4.4   Further Possible Module Splittings

For various reasons, further splittings of the modules of Figure 7.21(d) may be considered. Some possibilities are shown in Figures 7.22 and 7.23. For simplicity we consider only the interaction between one packet process (P1) and two links, namely its own link (link 1) and one other link (link 2).

Figure 7.22 illustrates some splittings which may be considered for a multiprocessor configuration. As was indicated in Section 7.4.2 the architecture of Figure 7.21(d) requires the packet processes to be in common memory and executable by any processor. What if there is no mechanism in the kernel, however, for a process to request the necessary transfer of allegiance? Or what if even the relatively small number of instruction-fetch bus cycles used up by the packet processes individually must be eliminated because the system must support a large number of links?

Then one solution is to split each packet process and to provide a small handover monitor as shown in Figure 7.22(a). The P11 and P12 processes may now be located in private memory with only a small handover monitor in common memory between them. The processes P11 and P12 are copies of P1 except with calls to the now inaccessible other DLC monitor replaced by calls to the handover monitor. The handover monitor itself simply implements a length one queue for handover of a buffer. This arrangement obviously violates Principle 3; some small amount of recoding and recompilation is required.

A solution which does not violate Principle 3 is to split each DLC monitor between two processors as shown in Figure 7.22(b). This somewhat awkward arrangement is easy to implement. The code of each monitor is split into a high and a low half which are located in the private memories of separate processors. All shared monitor data structures must be located in common memory. They are still protected by the monitor code, so their location presents no problems. However, an additional processor has been introduced. The arrangement would only make sense if this processor ran all link packet processes and the application processes as well. This is likely to be possible if most of the monitor complexity is in the low half, as may often be arranged with protocol monitors.

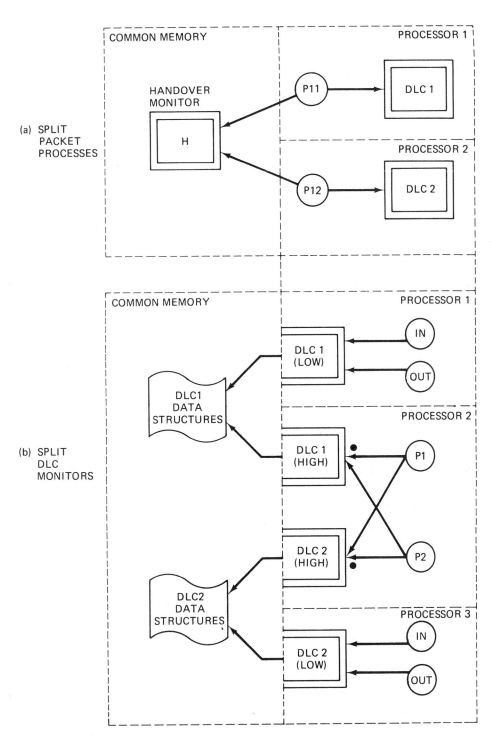

*Figure 7.22: Further Possible Module Splittings to Accommodate Multiprocessors*

Clearly a new principle is at work:

Principle 4

If all process and monitor instruction fetches must be strictly local to all processors (for efficiency or other reasons) then a split-process or split-monitor organization is required.

In our example design, we shall not invoke Principle 4 because it would unnecessarily complicate our simple example. Therefore no changes or additions are required to Figure 7.21(d).

Other possible splittings come to mind which at first glance may appear to be attractive but which on further examination are to be avoided. Some of these are illustrated in Figure 7.23. They all violate Principle 2.

In Figure 7.23(a) two monitors are defined to handle link operations, one to handle all incoming traffic and the other to handle all outgoing traffic. This approach has the spurious attraction of reducing interference between the two traffic streams. Unfortunately the two streams must interact because of the protocol requirements and therefore nothing is to be gained by the split. And simplicity and clarity are lost. The protocol FSMs must now be split between two monitors and inter-monitor calls must be introduced to pass on incoming protocol messages, as shown.

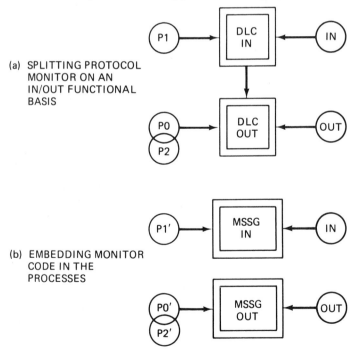

*Figure 7.23: Undesirable Module Splittings*

Figure 7.23(b) carries this idea even further by eliminating the protocol monitors entirely and replacing them with message monitors. The protocol FSMs must be implemented somehow by code in the processes. The message monitors simply hand over buffers.

Note only does this result in complex algorithms for the processes themselves, but it greatly reduces system modularity. Protocol changes now result in code changes in many processes and possibly in many places in each process.

### 7.4.5  Uniprocessor Efficiency

When outside constraints limit the number of processors to one, then efficiency of execution on a single processor will be important. Let us examine the efficiency of the various alternatives of Figure 7.21 from a uniprocessing viewpoint.

A monolithic monitor as in Figure 7.21(b) handling all links is inefficient on at least the following grounds:

— unnecessary mutual exclusions are enforced by the monitor in cases where calls are made which cannot interfere with each other because they reference different links.

— an extra level of indirect addressing is required to access link variables, which is inefficient on many microprocessors.

The only reason to combine functions which do not interact into a monolithic monitor is simply to share code. If code sharing is an important requirement, then the system may be structured as shown in Figure 7.24 so that the code to be shared is encapsulated in a reentrant subroutine module external to the particular processes or monitors involved. The monolithic monitor may then be partitioned as shown. However this approach will still have its own efficiency problems if reentrant code is inefficient for the particular processor or programming language employed. If the frequency of unnecessary mutual exclusions is low (as will usually be the case when the number of links is small), then no clear case can be made for the approproach of Figure 7.24(b) on the basis of uniprocessor efficiency alone.

Turning to the system of Figure 7.21(c), inter-monitor calls may result in monitor gate closure periods which are excessively long (even if deadlock is avoided). While one link monitor calls the other, both link monitor gates are closed and processes may be locked out unnecessarily.

And the "final choice" of Figure 7.21(d) is not without its faults. There are more processes in this design, leading to more process switching overhead and to the possibility of excessive monitor gate lockout resulting from processes being switched while executing monitor procedures. The latter effect may be eliminated by allowing a process to become non-switchable when in a monitor, at the cost of more switching complexity (and overhead). The switching overhead is a function of the kernel design and the hardware, neither of which may be yet decided at this level of design.

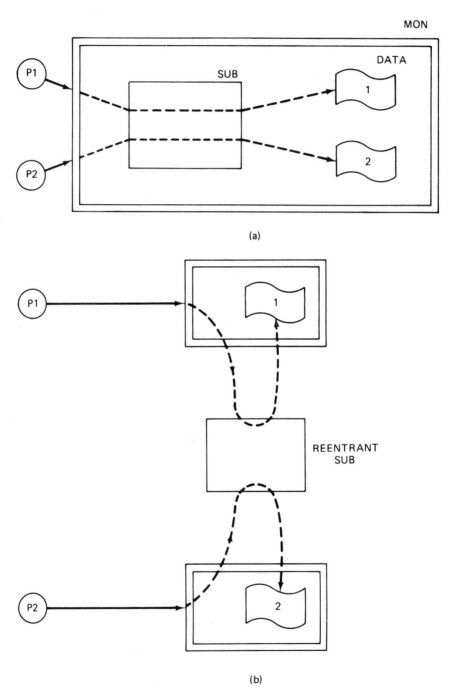

Figure 7.24: Monitors and Code Sharing

To summarize, there are a number of tradeoffs involved and the issues are not entirely clear-cut. Large monitors used to share code may provide unnecessarily slow monitor access and execution. Partitioning the monitors may speed up the direct monitor access but introduce new problems due to inter-monitor calls. And adding new processes to handle inter-monitor interfaces may introduce new delays due to higher process switching overheads and latencies; however, if the kernel and hardware are efficient, these delays may be small. More detailed design information is required to establish relative efficiency and this must be postponed to later design stages. If the issue is not clear-cut, the final arbiter is often the designer's personal taste.

We can only say at this stage that there appears to be no reason to reject the configuration of Figure 7.21(d) on the basis of uniprocessor efficiency.

### 7.4.6    Responsibilities of Processes

Other design principles may affect the "finding" of processes. For example, to make individual processes and their direct interactions as simple as possible, the following principle could be adopted:

Principle 5

Individual processes should not wait in multiple places for autonomous events nor should they manage multiple antonomous activities.

This is a very restrictive principle and is not always easy to follow. It applies to situations where multiple processes interact with each other. It should ensure that each process by itself is logically simple and easy to analyze for correctness. It also should simplify analysis for inter-process deadlocks, because the processes themselves are simple and their waiting patterns are clear and consistent. However, note that if too many processes result from the application of this principle, then their very numbers may be the cause of compensating additional complexity because of the difficulty of keeping track of indirect interactions. Finally it should ensure that concurrency is not compromised.

In cases where Principle 5 cannot be followed, the following principle may be substituted:

Principle 6

Where a process cannot avoid having responsibility for managing multiple autonomous activities, it should have one primary responsibility and use timed waits or polling to handle the others.

How do the designs of Figure 7.21 fare with respect to these principles? Consider Figure 7.21(d) first. Clearly the IN and OUT frame processes satisfy Principle 5. Each process is responsible for one external physical activity and has one place to wait for work (the DLC monitor in the case of the OUT process and the ISR level in the case of the IN process).

As for the packet processes, it depends on how their responsibilities are assigned. If these processes wait for acknowledgments of each forwarded datagram packet (as shown in Figure 7.25(a), then Principle 5 is violated. And the OUT processes are redundant (the sending process might as well do the work of the OUT process, since it can't do anything else). However, if the sending processes do not wait for acknowledgments, (as shown in Figure 7.25(b)), then Principle 6 is satisfied. Each packet transport process is then responsible for the primary activity of picking up packets on its own link with the other activities being clearly subsidiary. And the OUT processes are required. In the former case there are only two processes per link and concurrency is compromised. In the latter case there are three processes per link and concurrency is not compromised.

The application processes themselves satisfy Principle 5, at least from the limited view of them presented here.

Clearly, from this discussion, internal processes, such as the packet transport processes in Figure 7.21(d), have trouble satisfying Principle 5. The Principle can only be truly satisfied by the addition of new processes once again. Sufficient copies of each internal process may be provided (as shown in Figure 7.25(c)) so that each possible autonomous activity has its own copy. However, so many processes introduce additional complexities and inefficiencies. A particular problem is that of the absence of any guaranteed ordering of events between copies of processes, resulting in difficulty in analyzing for deadlocks (for example, due to buffer contention) and in a probable requirement for additional synchronizing mechanisms.

Finally, as shown in Figure 7.25(d), a higher level control process could perform the link checking functions. An advantage of this approach is that the control process could also check for link failures and initiate appropriate recovery actions.

Is it better, then, to attempt to avoid the problem of multiple responsibilities entirely by not introducing internal processes (as in the configurations of Figure 7.21(b) and 7.21(c)) in cases where they are not required for multiprocessor reasons? The answer appears to be no, because the same problems arise. Principle 5 cannot be satisfied for the application processes and Principle 6 can only be satisfied by requiring the application processes to use timed waits and polling for arriving data and acks.

For our simple example we shall choose a combination of Figures 7.25(a) and (d). For simplicity, the normal operation of the packet processes will follow the pattern of Figure 7.25(a). However, for error recovery a control process will be introduced as suggested by Figure 7.25(d).

### 7.4.7   Buffer Management

The next step is to refine further our chosen architecture of Figure 7.21(d) as modified by Figure 7.25(d) (by the addition of a control process) by considering how data buffering and buffer flow fit into this structure. Several approaches are possible.

Each link interrupt service routine (ISR) could own its own frame buffers (two each, for double buffering). The contents of these buffers could be copied into higher

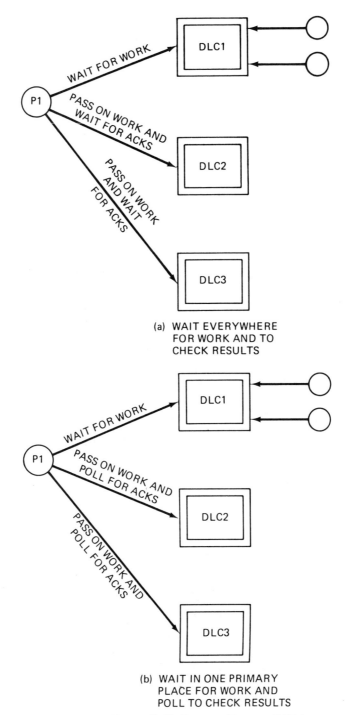

(a) WAIT EVERYWHERE
FOR WORK AND TO
CHECK RESULTS

(b) WAIT IN ONE PRIMARY
PLACE FOR WORK AND
POLL TO CHECK RESULTS

*Figure 7.25: Process Responsibilities*

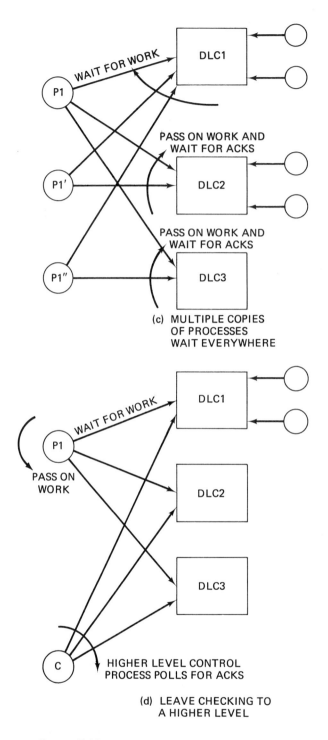

(c) MULTIPLE COPIES
OF PROCESSES
WAIT EVERYWHERE

(d) LEAVE CHECKING TO
A HIGHER LEVEL

*Figure 7.25: Process Responsibilities (Cont'd)*

level datagram buffers by the transport processes or the DLC monitor. Or, less wasteful of storage and copying time, a pool of generalized packet buffers suitable for both datagrams and frames could be shared by all processes. The latter approach will be used for the example system.

Assume there will be a single packet buffer pool monitor shared between all processes and all links. Now, two further architectural questions arise. Who will be allowed to access the buffer pool monitor? And how can the buffer flow be managed to accomplish the required functions, to avoid deadlock when insufficient buffers are available and to return buffers to the pool both during normal operation and after link restart? We now treat each of these questions in turn.

With respect to access to the buffer pool, consider Figure 7.26, which shows an access graph for a single link, displaying the possible multiprocessor configuration of the system. For purposes of this discussion, the buffers are assumed to be returned to the pool either by P0 and P1 (solid arrows in Figures 7.26(a) and (b)) or by DLC (dotted arrow in Figure 7.26(b)). We defer for the moment the possibility of a higher level control process doing the job. The link is assumed to be handled by a separate processor with the buffer pool in common memory. What is to be the nature of the hardware interface to the link processor? Is it to be a master/slave interface? In that case, calls from the link side of interface should be forbidden; in particular neither the frame processes nor the DLC monitor should be allowed to call the buffer pool monitor to acquire or release buffers, as shown in Figure 7.26(a). Or is it to be a multiple-master interface? In that case, calls from the link side may be allowed, as shown in Figure 7.26(b).

The multiple-master software configuration of Figure 7.26(b) implies a commitment to a multiple master hardware interface between processors, whereas the master-slave software configuration of Figure 7.26(a) may be implemented on any hardware configuration. In particular, a special purpose, very high performance slave processor could be used to implement data link control. It could even be placed on a single chip (chips which perform the functions of current SPAR protocols are commercially available). In such cases, a master/slave relationship is probably the only practical one.

As has been pointed out before, multiple-master hardware configurations are more difficult and expensive to implement than are master-slave configurations. However, there is a payoff in software modularity and simplicity, as we shall see.

Based on the above discussions, another design principle may now be enunciated:

Principle 7

Where it can be foreseen that the software level on one side of an interface may be eventually implemented on a special-purpose, high performance, slave device, then that level should be designed in software as a slave level.

This principle is quite restrictive and can result in a certain awkwardness in buffer flow management. For example, initial buffers for the ISR level must be passed by the

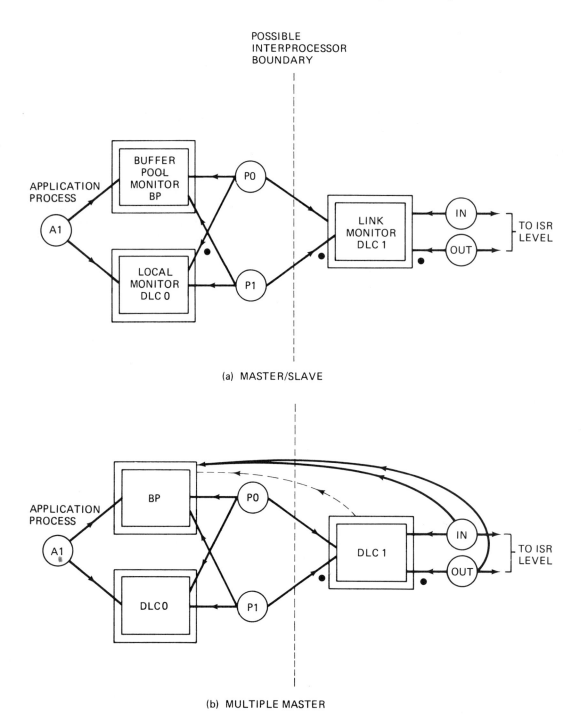

(a) MASTER/SLAVE

(b) MULTIPLE MASTER

*Figure 7.26: A Possible Multiprocessor View of the Software Architecture*

packet processes via the DLC monitor in the configuration of Figure 7.26(a); there is nowhere else they can come from. Logical clarity of the DLC monitor is thus compromised.

For our example system we shall not invoke Principle 7 and shall therefore proceed with the configuration of Figure 7.26(b). This configuration is both logically simpler and more modular than the master-slave one. It is simpler because awkward buffer handover mechanisms do not have to be installed in the packet processes and DLC monitors to provide initial buffers to the ISR level. And it is more modular because no commitment has been made in software to any particular location of the interprocessor interface, provided the hardware is organized in a multiple master configuration.

Having chosen the form of the access graph, we may now proceed with defining the buffer flow patterns. Figure 7.27 is a data flow diagram which shows possible flow patterns for the access graph of Figure 7.26(b). In this figure, the arrows representing flows begin and terminate on dots which represent waiting positions for buffers.

In Figure 7.27(a), an empty buffer is picked up, filled by application process A1 and then passed on to the appropriate link; after reception has been acknowledged, the process receiving the acknowledgement (in this case P0) returns the buffer to the pool. Alternatively, as shown by the dotted lines, the DLC monitor could return the buffer.

In Figure 7.27(b) the IN process of Link 1 receives a filled buffer from the link. It also keeps the link primed with empty buffers via the "replenish" path shown in the figure. The subsidiary loop at the link level shows that arriving frames with framing errors are simply discarded and not passed on to the IN process. The buffer is then forwarded to another external link (link 2 here) by P1, who returns the buffer to the pool after acknowledgement is received. Alternatively the buffer could be returned by the DLC monitor as shown by the dotted lines.

Figure 7.27(c) shows a similar cycle except that the forwarding is to the local link and return of the buffer to the pool is by the application process receiving it.

Consider further the question of who should return transmitted buffers to the pool after their reception has been acknowledged. Three alternatives are open:

(a) The processes which transport packets between DLC monitors (as shown by the solid lines in Figures 7.26 and 7.27);

(b) The DLC monitors themselves (as shown by the dotted lines in Figures 7.26 and 7.27);

(c) A separate, higher level control process (not shown in Figures 7.26 or 7.27 but suggested in Section 7.4.6).

Alternatives (a) and (c) are both compatible with a master/slave hardware interface. Of the two, alternative (c) appears preferable not only for the reasons discussed in Section 7.4.6 but also because of improved system throughput (while a transport process is waiting for an ack on one link it can't transport new packets to other links).

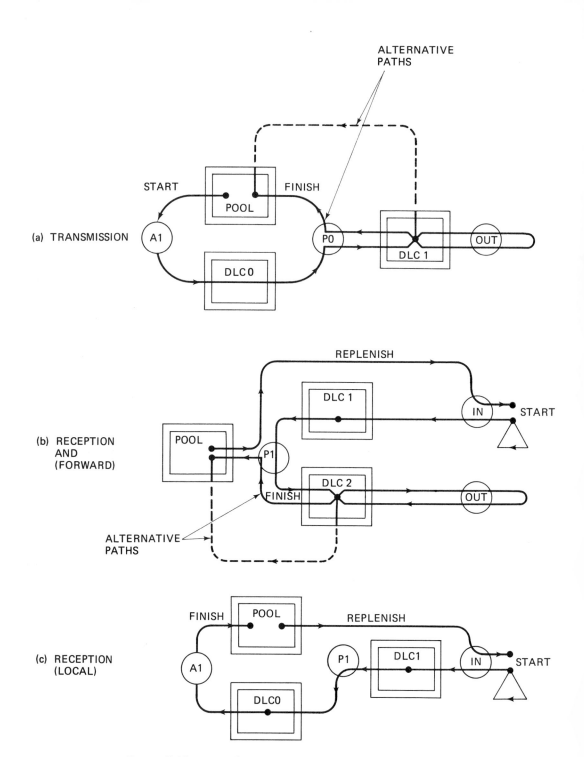

*Figure 7.27: Data Flow Diagram Showing Buffer Flow Possibilities*

Alternative (b) on the other hand has all the advantages of alternative (c) provided a multiple master hardware configuration is available.

For our simple example we shall opt for alternative (b).

We defer for the moment to the question of how deadlock due to a circular wait for buffers can be avoided when insufficient buffers are available. This question will be considered in Section 7.5 where actual monitor interfaces and process algorithms are discussed.

## 7.4.8   System Restart and Control

On power-up a node will attempt to start up all its links. On link restart only the failed link will be started up. In the latter, case some active processes may have to modify their activities and active buffers may have to be returned to the pool. For uniformity and modularity, it is advantageous to treat all startup situations with full generality. Thus power-up should be handled identically to link restart. This implies that the control of startup should reside in the active elements of the system, namely the processes, rather than in the passive elements, namely the power-up software and the monitors. Another principle is at work:

Principle 8

Control of startup after power-up and of restart after failure should reside in the active components of the system, namely the processes.

Therefore the system monitors should provide calls to set link status to LINK DOWN, to read link status and to initiate link restart actions to bring the link status to LINK UP. Furthermore, these actions do not occur only on an isolated link basis. Because of the inter-link transfer activities of the system, failure of one link may in general require high-level recovery actions associated with another link.

Therefore the link status and restart activities are best performed by a separate control process which can coordinate these activities for the entire node. This process is the logical one to trigger the end-to-end recovery actions which are outside the scope of our example.

Having decided on the requirement for a control process, we may now return to the question raised in Section 7.4.7 of who should return the failed link's buffers to the pool, the control process or the DLC monitor.

It is difficult to state a specific principle to invoke in assisting with this sort of decision. However, undaunted we offer the following general principle:

Principle 9 (the KISS or "Keep It Simple, Stupid", principle)

Logical clarity and modularity of a concurrent system are enhanced by keeping the functions of individual processes simple, distinct and logically layered; and by encapsulating logically complex lower level functions in passive modules.

Principle 9, properly applied, should result in a system structure which can be explained to and understood by almost anyone, even by managers and salesmen! We may say that modules should have clearly recognized "personalities". If a designer can achieve this state of clarity, then he can be sure that the implementors and maintainers of the system will have little trouble with it.

Principle 9, applied to the control process implies that its functions should be confined to the logically high level ones of checking-link status, and requesting that recovery action be initiated. Buffer return management is clearly a lower level function which is best encapsulated in a lower level module. The DLC module is the logical one; buffer returns by that module are then logically and uniformly triggered by internal module status changes resulting from both normal link operation and special requests by the control process.

A timeout mechanism is required for link startup (and in general for link operation). A clock ISR is required. And since ISRs cannot call monitors, this ISR must wake up a timeout (T) process via a semaphore; the T process then calls the DLC monitors.

Finally, we must decide how the control process is to synchronize its activities with respect to the other activities of the system. It needs a central place to wait for something to happen. An appropriate approach is to have it wait in a central control monitor for requests from other system levels. To provide clocked wakeup of the control process, the timeout process must call the control monitor as well as the DLC monitors. On awakening, the control process polls all DLC monitors for link status and requests recovery actions be initiated if necessary.

### 7.4.9 Software Pipelining

An architecture which suggests software pipelining is shown in Figure 7.28(a). For example, A could be A1 of Figure 7.26, B could be PO and C could be OUT. Software pipelining occurs if A hands over work to B and then continues with its own work. Similarly, B picks up work from A, processes it, hands over consequential work to C and then proceeds with its own work. A, B and C all are intended to operate concurrently, at least in principle. Instances of this type of architecture are prevalent in communications software and transaction processing.

However, in practice, constraints on concurrency in this pipelined architecture may occur for two reasons:

— The preliminary design studies may have overestimated the requirement for concurrency; logical detailed design development may show that A hands over work to B and then waits for it to be done and so on.

— Error recovery problems may dictate that pipelining be limited, even where it appears desirable; because work on different steps of many transactions may be progressing simultaneously, recovery from failure in any step of one transaction may become complex.

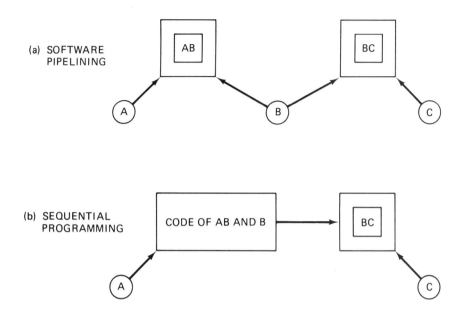

*Figure 7.28: Software Pipelining versus Sequential Programming*

In such cases, there may be no need for certain processes. Then these processes may be converted to subroutines when they are programmed. For example, if A waits for B to finish, B may be replaced by a subroutine as shown in the Figure 7.28(b).

Any design should be carefully examined for instances of apparent software pipelining which are in fact sequential programming. Then a decision must be made on whether or not to leave the structure as is or to make the adjustment suggested by Figure 7.28(b). The disadvantage of leaving it is that there are unnecessary asynchronous activities taking place which reduce efficiency and which may increase logical complexity of the system as a whole. One advantage is that the way is left open for future real pipelining if it can be foreseen as a requirement. Another is that processes themselves are simpler.

## 7.4.10   Final System Architecture

We have now identified all the major communications system modules and defined their functions and interfaces informally. The only significant remaining objects to be included in the access graph are the shared data structures required by the communication links. The principal among them are the routing tables, which provide the "next link" identification for forwarding a datagram packet to an end address. The tables are static (read only) because dynamic updating was specifically excluded from the functional requirements. Accordingly, there are no coordination problems to be solved and design of the tables can be postponed to a more detailed level.

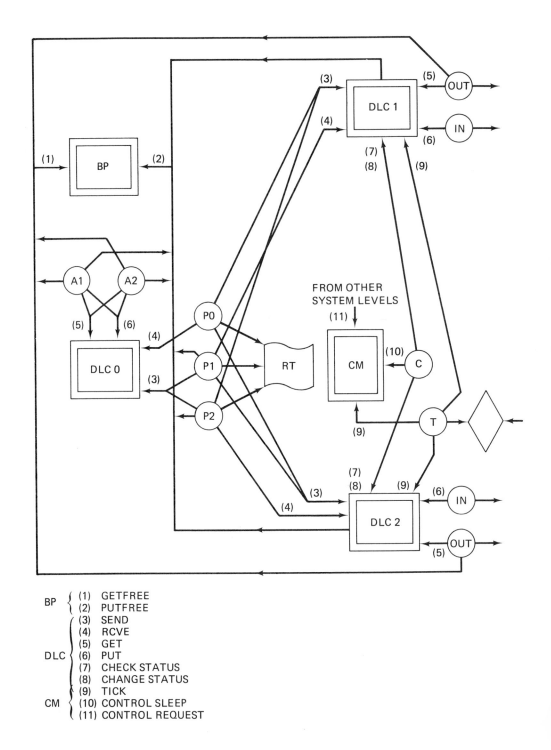

BP $\left\{\begin{array}{ll}(1) & \text{GETFREE} \\ (2) & \text{PUTFREE}\end{array}\right.$

DLC $\left\{\begin{array}{ll}(3) & \text{SEND} \\ (4) & \text{RCVE} \\ (5) & \text{GET} \\ (6) & \text{PUT} \\ (7) & \text{CHECK STATUS} \\ (8) & \text{CHANGE STATUS} \\ (9) & \text{TICK}\end{array}\right.$

CM $\left\{\begin{array}{ll}(10) & \text{CONTROL SLEEP} \\ (11) & \text{CONTROL REQUEST}\end{array}\right.$

*Figure 7.29: The Final Access Graph*

The final data flow diagram is as shown in Figure 7.27 with buffer return to the pool via the dotted lines from the DLC monitors.

The final access graph is shown in Figure 7.29 for two links.

Before beginning detailed design, one issue must be resolved: to time-slice or not to time-slice? Processes in a rigidly designed, special purpose system may not need to be time sliced. The designers may have enough control over the system to guarantee that natural preemption (waiting on a semaphore or blocking in a monitor) occurs often enough to ensure equitable access to processors. However, clock-controlled time slicing may be required for evolving and user-programmed systems in which this cannot be guaranteed. And it frees the designer from worrying at design time about total process execution time. We shall assume clock-controlled preemption will be provided. Further discussion of the clock preemption issue follows in Chapter 9.

## 7.5   Algorithmic Design of the Major Active Modules

Our task now is to develop the overall algorithmic features of the architecture chosen in Section 7.4 (Figure 7.29), including functions and interfaces of the monitors, functions and actions of the processes and overall system operation and control. The emphasis is on the active modules of the system; only the external interfaces of the passive modules are involved.

Section 7.5.1 defines the external interface of the DLC monitor. Section 7.5.2 discusses buffer pool management by the processes to avoid deadlock. Then Section 7.5.3 develops algorithms for the major processes.

### 7.5.1   External Specification of the Data Link Monitor

Figure 7.29 shows that the DLC monitor must provide four data-oriented entry procedures SEND, RCVE, PUT and GET; two recovery-oriented entry procedures CHECK STATUS and CHANGE STATUS; and one timeout-oriented entry procedure TICK. Figure 7.30 illustrates the entry procedures and shows how they are functionally related (dotted lines).

SEND and RCVE are for use by high level processes to send and receive datagram packets. SEND may be called by any high level process. SEND always suspends each caller until the frame is acknowledged by the receiving end of the link; if more than one caller is suspended, they are served in FIFO order. RCVE only suspends its callers if there are no packets to receive; suspended callers are served in FIFO order when frames arrive. However, RCVE is normally called only by one process, namely the packet transport process reserved for listening for input on that link.

GET and PUT are for use by low level frame processes; each is called by only one process and the caller is never suspended. For the example system, GET always provides the caller with a frame for transmission, either a data frame or, if no data frames are waiting, an explicit ACK frame which will provide acknowledgement for

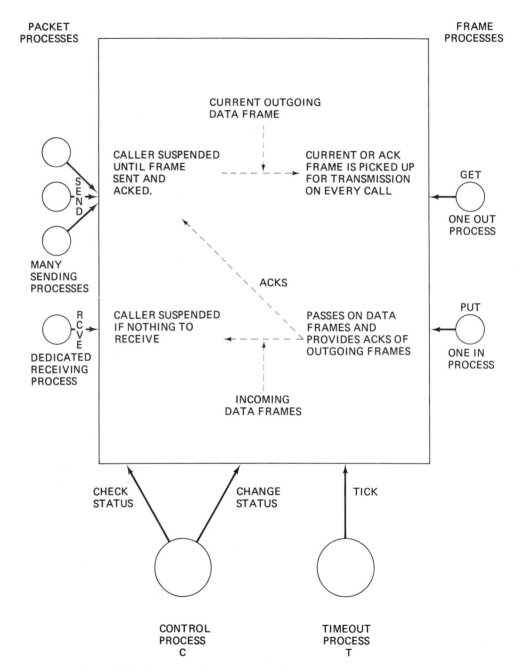

*Figure 7.30: External Specification of the Data Link Control Monitor*

received frames. PUT always takes the incoming frame from the caller, and performs the twin functions of passing on data frames for pickup by waiting callers of RCVE and of passing on ACK information for waiting callers of SEND.

CHECKSTATUS is a no-suspension call for use by the control process to check the current status of the link. CHANGESTATUS is for use by the control process to request shutdown and restart of the link.

TICK provides internal timer updates for protocol timeouts. It is called only by the timeout process T.

Note that the interface to the DLC monitor as described above is quite general in nature but that certain detailed simplifications have been made for pedagocial purposes, as follows. Instead of assuming an internal DLC monitor timeout on unacknowledged data frames, we assume that the OUT process repetitively calls GET on a cyclic basis determined by its own internal logic to perform repetitive retransmissions. It is only a small additional complication to provide a timer recovery mechanism driven by the TICK call. As it stands, the TICK call is required only for startup protocol timeouts. And instead of providing a multi-buffered transmission mechanism in the DLC monitor, each process calling SEND must wait until its packet is acked before leaving the monitor. Multiple calls to SEND are allowed but callers are suspended in the monitor until all previous callers have been completely served. It is another small additional complication to provide multiple buffering. However, the combination of timer recovery and multiple buffering adds significant internal complexity to the DLC monitor.

### 7.5.2 Buffer Deadlock Avoidance

Before proceeding with algorithmic design of the processes, it is important to formulate rules for avoiding deadlock due to circular waits for buffers.

Figure 7.31 shows the general nature of the relationship between the communications package, the buffer pool, and application processes which access the communications interface. Accesses to the pool are made from within the communications package, both for return of transmitted buffers and for acquisition of empty ones for reception. Similarly, the processes above the communications interface acquire buffers from the pool for filling before transmission and return buffers received from the interface to the pool after using their contents. Two representative life cycles are shown in the figure by dotted lines. Process A1 acquires an empty buffer from the pool, copies a portion of the message area into it, appends the datagram address fields and sends the buffer which eventually trickles back into the pool from the communications interface. Process A2, on the other hand, receives buffers which have been picked up from the pool and filled by the communications interface, copies them into the message area and then releases them to the pool. A prime concern is to ensure that processes cannot deadlock due to a circular wait for buffers.

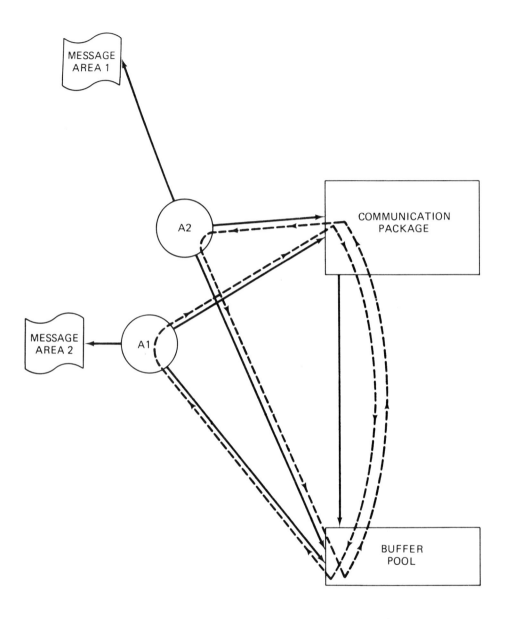

A2   IS RECEIVING PACKETS AND DEPACKETIZING

A1   IS PACKETIZING AND SENDING

DOTTED LINES SHOW BUFFER LIFE CYCLES.

*Figure 7.31: Communications Buffer Flow*

This section prescribes simple rules for using the buffer pool procedures GET FREE and PUT FREE to ensure that no deadlock can occur due to pool accesses alone. Deadlock is possible only if:

— a process requests more than one buffer, one at a time, to obtain the number required to complete its function, and

— the pool is smaller than the total simultaneous maximum demand by all processes.

If the pool is small, then the first condition must be prevented; that is, processes must not hold more than one buffer at a time.

First we note that a buffer given by a process to the communication package by SEND inevitably finds its way back to the pool. (This must be guaranteed in the design of the communications package). In effect, then, transferring a buffer by SEND is equivalent to returning it to the pool. This leads to:

Rule 1:

A process sending a message by repeated calls to GET$FREE and SEND shall never hold more than one buffer at a time; that is, calls to GET$FREE and SEND strictly alternate.

Next we consider reception. If processes were allowed to hold a buffer obtained through RCVE, and to call RCVE again before returning the first buffer by PUT$FREE, then a small pool could be emptied. Buffers would be held by application processes outside of the pool and the communication package. Hence:

Rule 2:

A process receiving a message shall strictly alternate calls to RCVE and PUT$FREE.

Adherence to this rule means that buffers obtained from the pool by the communciation package inevitably find their way back to the pool in finite time.

A certain minimum number of buffers is required to keep the reception side oiled. A pool larger than this minimum can be operated by an arbitrary number of processes without fear of deadlock. Increasing the size of the pool simply makes the system run more smoothly.

We see, therefore, that the pool needs no protection other than the normal mutual exclusion provided by a monitor. No bulky and time consuming deadlock avoidance algorithm is required in the pool if processes adhere to the two rules and the communications interface guarantees eventual return of buffers to the pool.

Thus, it is possible to avoid deadlocks due to buffer pool accesses alone. This does not guarantee that the system as a whole will function without deadlocks, however. Deadlocks are still possible due to competition for other resources.

In order to keep life simple we can adopt the following:

Rule 3:

A process shall access the pool and the communications interface only after all its other resource ownership negotiations are complete, and shall not attempt to claim another resource while it holds any buffers.

Even prevention of deadlock within the node as a whole does not necessarily prevent deadlock when two nodes are connected by a network link. These problems tend to be subtle; until more is known about the future applications of communicating systems it will be difficult to formulate easy, general rules.

### 7.5.3 Algorithmic Operation of the Processes

Because all the complicated details are buried in the monitors, the processes are extremely simple. The manipulation of data structures and the interaction with other processes is hidden in the monitor procedures. All the processes must do is provide correct sequencing and parameters of monitor calls.

Each link has its own dedicated packet transport process which is responsible for picking up packets from its own DLC monitor and forwarding them to the appropriate next DLC monitor (recall that local application processes are served by a degenerate, local DLC monitor which looks to the local datagram transport process just like a conventional link monitor). The life of a transport process is spent in a basic monotonous cycle of two procedure calls: RCVE from its own monitor and SEND to another one.

At the frame level, each link has a pair of transport processes, an OUT process to pick up frames for transmission and an IN process to receive frames. Each of these processes is similarly trapped in an endless cycle. The OUT process calls GET and then delivers the frame to the link ISR (interrupt service routine) level. The IN process picks up a frame from the link ISR level, calls PUT and then calls GETFREE for an empty buffer to replenish the ISR level's buffer supply.

The control process C spends most of its life asleep in the CM monitor. Periodically or when requested by other system layers (not described here) it polls the CHECKSTATUS routines of all DLC monitors and calls CHANGESTATUS to request links be shutdown and restarted, as appropriate. In a more sophisticated system it might also trigger the recovery mechanism of an end-to-end protocol which would ensure that datagrams forwarded via failed links are not lost. However, in the simple example system, no such mechanism is present and such datagrams will be lost. All that the control process can ensure is that individual links are restarted when required.

The timeout process T has an extremely simple function: to provide periodic TICK calls to the CM and DLC monitors.

At the application level, datagram packets are deposited and picked up by the local application processes from the local DLC monitor by special high level versions of the

PUT and GET calls which pass datagrams untouched through the monitor.

Each process has a "home" where it waits on startup or restart, as follows:

— Each packet transport process calls RCVE in its own DLC monitor where it waits for a packet to arrive over the link (or to be passed on by an application process in the case of the local DLC monitor).

— Each link OUT process calls GET in its own DLC monitor where it waits until the link is up (recall that it never waits during normal operations).

— Each link IN process waits at its link ISR level for an arriving frame. Note that at this point the link ISRs must have sufficient empty buffers in their possession.

— The C (control) process waits in the CM monitor for the system to tell it to do something.

— The T (timeout) process waits on a private semaphore.

Informal outlines of the code for all processes except IN and OUT are shown in Figure 7.32. The IN and OUT processes are described in Section 7.7 and Figure 7.38.

```
        .
        .
        .
    DLC$GET (BUFADDR);
    COPY BUFFER INTO MSG AREA;
    BP$PUT$FREE (BUFADDR);
        .
        .
        .
    BP$GET$FREE (BUFADDR);
    COPY MSG INTO BUFFER;
    DLC$PUT (BUFADDR);
        .
        .
        .
```

*Figure 7.32(a): Application Process Informal Code*

```
WHILE TRUE DO
BEGIN
  DLC$RCVE (BUFFADDR);
  LOOK UP ROUTING TABLE FOR LINK ;
  CASE LINK OF
        .
        .
        .
        .
    TWO:  DLC2$SEND (BUFADDR);
        .
        .
        .

    END

END
```

*Figure 7.32(b): Packet Process Informal Code*

```
WHILE TRUE DO
BEGIN
        CONTROL SLEEP (EVENT);
        CASE EVENT OF
        POLL:
        . . .


          RESTART ALL:
            . . .
        END
```

*Figure 7.32(c): Control Process Informal Code*

## 7.6  Algorithmic Design of a Passive Module: The Data Link Control Monitor for Link Operation

Here follows a step-by-step development of the detailed design of the DLC monitor.

From Figure 7.30 which shows interaction between the monitor entry procedures, as seen externally, and Section 7.3 which describes the link protocol, an internal access graph and informal pseudo-code are developed for the monitor entry procedures. Then a more formal description of the monitor procedures is presented. The intent is to display a typical step-by-step development of a monitor, instead of simply presenting a final solution.

Three major components of the monitor are as follows:

— protocol state variables and transition rules (described in Section 7.3.3);

— places for processes to sleep (condition variables) and strategies for awakening them (described below); and

— queues for packet buffers (described below).

For illustration purposes we shall here develop the DLC monitor to implement the link operation protocol only. Accordingly, the state variables of Section 7.3.3 are reduced to a single boolean variable BUSY, which is false when the link OP protocol is IDLE and true when the protocol is BUSY.

From Figure 7.30, the following condition variables and queues are required for link operation:

1. A place for processes calling SEND to wait while the link is busy; this is the LINKAVAIL condition variable.

2. A place for the sending process currently being served to wait for acknowledgement of its frame; this is the CONFIRM condition variable.

3. A place for processes calling RCVE to wait for arriving frames; this is the ARRIVAL condition variable.

4. Queues for incoming and outgoing frames; these are called SENDTHIS and RCVETHIS.

There are two possible strategies for awakening processes in monitors as illustrated in Figure 7.33:

Selective:

1. Processes doing the awakening must have the intelligence to know who to awaken (this strategy is most appropriate for normal processing functions such as delivering packets (Figure 7.33(a)).

Universal:

2. All processes are awakened: Processes which are awakened must have the intelligence to decide whether to do anything or to go back to sleep. This strategy is most appropriate for error conditions, status changes, etc., where the process doing the awakening only reports the condition and should not have to know what to do about it (Figure 7.33(b)).

Informal pseudo-code for the monitor entry procedures SEND, RCVE, GET and PUT is now sketched based on the internal access graph of Figure 7.34, which employs the selective awakening strategy.

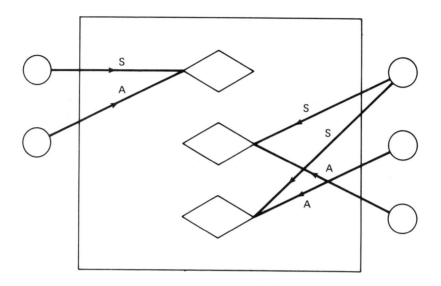

*Figure 7.33(a): Selective Awakening Strategy*

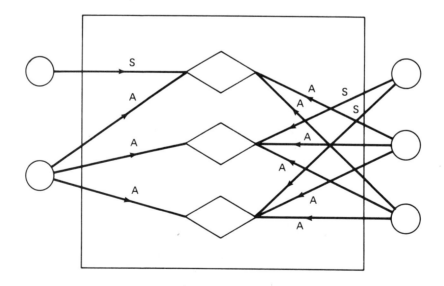

*Figure 7.33(b): Universal Awakening Strategy*

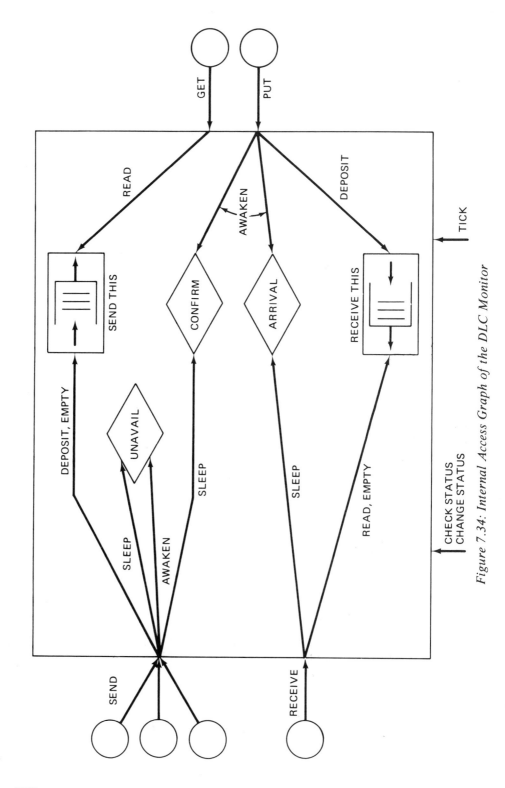

Figure 7.34: Internal Access Graph of the DLC Monitor

Figure 7.35(a) gives the SEND procedure. In line 3, the link is busy until the current frame, if one is active, is acked. Processes are suspended without queueing their packet buffers explicitly; at this point, the buffer address will be in the sleeping process' stack where it was put by the reentrant monitor entry procedure.

In lines 5 and 6, only those header fields are filled which are known unambiguously at the time. Filling in of the receive sequence number is postponed until the moment of frame pickup by the frame OUT process calling GET, because in the meantime further frames could arrive requiring acknowledgement. The remainder of the figure is self-explanatory.

Figures 7.35(b) and 7.35(c) provide the RCVE and GET procedures; they are self-explanatory.

Figure 7.35(d) provides the PUT procedure. Observe that this is the procedure that is in a position to raise the alarm on protocol failures shown by inconsistent count fields in arriving frames (lines 5 and 11).

```
 1. Procedure send (datagram address);
 2. Begin enter;
 3. While link is busy do sleep (linkavail);
 4. Mark link busy;
 5. Fill frame header fields for frame kind and send sequence
    number;
 6. (Note – leave receive sequence number for the GET procedure);
 7. Put incomplete frame in send this and mark send this full;
 8. Increment send sequence number;
 9. While frame remains unacked do sleep (confirm);
10. Mark link not busy;
11. Mark sendthis empty and return buffer to pool;
12. Awaken (link avail);
13. Exit end;
```

*Figure 7.35(a): DLC Monitor SEND Informal Pseudo Code*

```
 1. Procedure rcve (datagram address);
 2. Begin enter;
 3. If no datagram to pickup then sleep (arrival);
 4. Pick up the datagram from rcvethis;
 5. Exit end;
```

*Figure 7.35(b): DLC Monitor RCVE Informal Pseudo Code*

```
1. Procedure get (datagram address);
2. Begin enter;
3. If send this is not empty then pick up datagram address from
   send this;
4. (but do not mark send this empty)
5. Else frame address is of standard ack frame;
6. Fill receive sequence number field of frame header with
   number next expected frame;
7. Exit end;
```

(note that this procedure will keep getting the same frame to send until sendthis is marked empty by the send procedure after confirmation of an ack)

*Figure 7.35(c): DLC Monitor GET Informal Pseudo Code*

```
 1. procedure put (datagram address);
 2. begin enter;
 3. with the incoming frame do;
 4. begin
 5. if the frame contains a new acknowledgement;
 6. then begin
 7. save ack count;
 8. awaken (confirm);
 9. end;
10. if frame is an information frame
11. and it is the next expected frame
12. and rcvethis is empty
13. then begin
14. increment next expected sequence number;
15. put datagram in rcvethis;
16. awaken (arrival);
17. end
18. else return buffer to pool end;
19. exit end;
```

*Figure 7.35(d): DLC Monitor PUT Informal Pseudo Code*

From the link operation protocol requirements of Section 7.3 we see the need for the following internal protocol variables in the monitor (recall that the protocol requires that each frame has associated send and receive sequence numbers, here called ns and nr, respectively):

1. vs — the next send sequence number (ns) to be added to the next outgoing frame

2. vr — the next send sequence number (ns) expected in the next incoming frame (vr was called NESN in Section 7.3); this number is sent in the nr field of all outgoing frames to acknowledge all incoming frames with numbers less than vr; when an incoming frame with ns = vr is received then vr is incremented.

3. lastnr — the last receive sequence number (nr) in the last incoming frame; this number minus one is the number of the last local frame which was acknowledged by the other end of the link; further nr fields in incoming packets are ignored until they differ from lastnr.

Using the variables vs, vr, lastnr, we can now fill in the appropriate lines of the informal pseudo-code of Figure 7.35, as follows:

```
Send: line 5 _ frame header := vs
      line 8 _ increment vs
      line 9 _ current frame remains unacked while the number of the
      last local frame acknowledged by the other end of the link
      (lastnr-1) is not equal to the number of the last local frame
      sent (vs-1); i.e., while lastnr = vs
Get: line 6 _ outgoing frame header := vr
Put: line 5 _ incoming frame contains a new acknowledgement of nr =
lastnr
      line 7 _ lastnr := nr
      line 11 _ incoming frame is the next expected frame if its send
      sequence count equals vr; i.e., if ns = vr
```

Figure 7.36 summarizes the final DLC operation monitor in a more concise and formal notation. Note that all buffers are assumed to be in frame format. Note also that discarded buffers are returned to the buffer pool by the PUT procedure A new procedure, DLCINIT, has been introduced, which initializes internal monitor variables. However, note that the figure does not include the link startup protocol.

```
type byte = 0..255;
type frame = record kind: (info, ack, start, startack);
              nr: byte; ns: byte;
              data: array [byte] of byte end; ↑ frame;

var busy: boolean;
    lastnr, vs, vr: byte; sendthis, rcvethis:
    linkavailable, confirm, arrival: condition;
    ackskeleton: frame;

procedure send (sendptr: ↑ frame)
    begin enter;
    while busy do sleep (linkavailable);
    busy := true;
    sendptr↑.kind := info; sendptr↑.ns := vs;
    sendthis := sendptr; /*queue for transmission*/
    vs := (vs + 1) mod 2;                        /*for frame ack*/
    while lastnr <> vs do sleep (confirm); /*wait until frame
    acknowledged*/
    busy := false; sendthis := nil;
    awaken (linkavailable); exit end; /*send*/

procedure rcve (var rcveptr: ↑ frame);
    begin enter;
    while rcvethis = nil do sleep (arrival);
    rcveptr := rcvethis; rcvethis := nil; exit end; /*rcve */

procedure get (var outptr: ↑ frame)
    begin enter;
    if sendthis <> nil then outptr := sendthis /*info frame*/
    else outptr := ackskeleton; /*ack frame */
    outptr ↑.nr := vr; exit end; /*get*/

procedure put (inptr: ↑ frame);
    begin enter;
    with inptr↑ do begin
        if nr <> lastnr then begin
            lastnr := nr;
            awaken (confirm) end;
        if (kind = info) and (ns = vr) and (rcvethis = nil) then begin
            vr := (vr + 1) mod 2;
            rcvethis := inptr
            awaken (arrival) end
        else putfree (inptr) end; /*discard frame*/ exit end; /*put*/

procedure DLCinit;
    begin;
    busy := false
    sendthis := nil; rcvethis := nil;
    lastnr := 0; vs := 0; vr := 0;
    ackskeleton.kind := ack; end; /*DLCinit*/
```

*Figure 7.36: The Final DLC Monitor*

## 7.7   Frame Transmission/Reception

Previous sections have identified some components and functions of the frame transmission/reception level, namely the requirement for frame IN and OUT processes, the nature of their interfaces to the DLC monitor and the pattern of buffer flow through the frame level. Previous sections have also provided examples of stepwise, evolutionary development of software architectures and algorithms. The purpose of the present section is to present a fairly complete realistic design for the entire frame level below the DLC monitor. Variants of this design have been used in the authors' laboratory in a number of real multimicroprocessor systems implementing HDLC and X.25 protocols. To this level, it does not matter whether the higher levels implement the simple protocol described earlier or a much more complex protocol such as HDLC. This section will provide the reader with an understanding of how the ideas of the previous sections can be married to real hardware.

A data link control procedure must handle framing, transparency and error checking. These have the following meaning:

— *framing:*  the method of delimiting frames (transmission blocks) usually with special bit or character sequences.

— *transparency:*  the method of allowing unrestricted binary data without conflicting with the framing technique.

— *error checking:*  the method of detecting transmission errors so that only error-free frames need be processed.

These functions may be advantageously performed in modules separate from the main frame level (HDLC) monitor as shown in Figure 7.37. Several advantages derive from this approach:

— The design of the frame level DLC monitor is simplified

— Alternative frame structures can be used, for example the transparent BSC frame structure with HDLC procedures. Use of BSC framing allows asynchronous transmission over local links and therefore cheap data sets.

— These functions may be easily implemented using special chips, so that when convenient, both modules can be replaced by hardware. Most of the CPU time of an exclusively software implementation would be spent in these two modules.

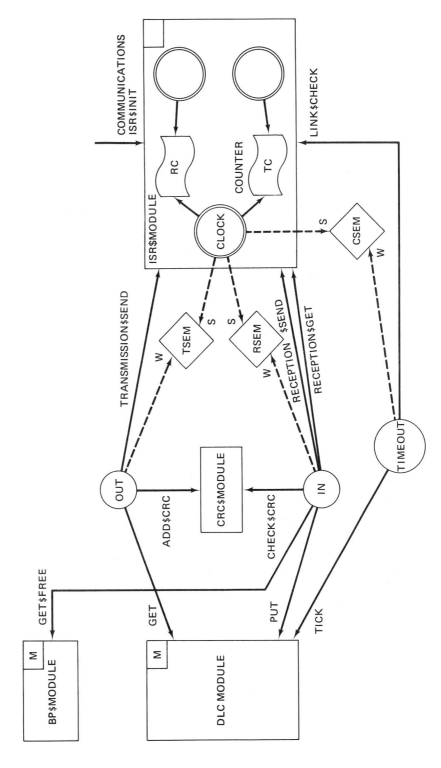

Figure 7.37: The ISR and CRC Modules

### 7.7.1    Flow of Buffers between the ISR and DLC Modules

Before proceeding further, the reader should refer back to Figure 7.27 which shows the migration of buffers in the example system, including the flow between the ISR and DLC modules. On the transmission side, process OUT gets a buffer from the DLC monitor and places it on the ISR's transmit queue. It then waits for transmission to be completed before returning to the higher level. Transmitted buffers are returned to the free pool at the higher level when an acknowledgement is received as described earlier. On the receive side, an empty buffer is obtained from the buffer pool monitor and placed on the receive queue from which it is retrieved when full. It is then subjected to the CRC check; if it fails the check it is recycled to the ISR module immediately, otherwise it is passed to the DLC module. Operation is clearly single-buffered.

However, in general, transmission and reception at the ISR level must be multiple buffered, for efficiency. For example, the transmission ISR can in general use up to three buffers:

— one buffer to hold the frame just transmitted;

— a second buffer to hold the current frame being transmitted;

— and a third buffer to hold the next frame to be transmitted.

A similar trio of buffers can be identified for reception. In general, the design of the ISR module should provide sufficient space to queue this many buffers. This space overhead is not onerous if pointers only are queued. The actual degree of multiple buffering may then be determined by the IN and OUT processes, as will be shown.

### 7.7.2    External Specification of the ISR Module

The ISR module of Figure 7.37 provides multi-buffered frame transmission and reception services with framing and transparency functions included. It is the "device driver" for the link. Multi-buffering is provided as a general feature; it is not required for the DLC module of Section 7.6. Although interrupts are used to move bytes between the frame buffer and the USART, interaction with the transport processes IN (interrupt to frame) and OUT (frame to interrupt) is by means of the procedure calls and semaphores listed below.

```
COMMUNICATION$ISR$INIT
```

A procedure used by the startup software which clears all data of the module and initializes the USART to the correct mode and parameters.

```
TRANSMISSION$SEND(PTR : ↑ BUFFER)
```

A procedure used by the OUT process to queue the specified buffer for transmission. If the transmission ISR is idle at the time of the call then it is restarted. In general several such calls may be made in advance of complete transmission of the current buffer (as indicated by the semaphore TSEM); the number of such calls is limited by the internal queue size (typically 3).

### RECEPTION$SEND(PTR : ↑ BUFFER)

A procedure used by the IN process to provide empty buffers; it queues the specified empty buffer for reception and subsequent pickup by RECEPTION$GET. In general several such buffers may be queued ahead; the number is limited by the internal queue size (typically 3). Queueing only one ahead ("single buffering") runs the risk of occasional data loss and consequent inefficiency due to retransmissions.

### RECEPTION$GET (VAR BUFADDR : ↑ BUFFER)

A procedure returning an address value used to obtain a filled buffer from the reception ISR. The call must be preceded by a wait on the semaphore RSEM to ensure that the buffer is ready. The buffer may be corrupted by transmission errors, but has at least satisfied the framing and transparency functions.

### LINK$CHECK

A procedure called by the TIMEOUT process to release buffers if the data set fails or is not ready. If the DSR lead on the EIA interface is not asserted, then the transmit queue is cleared and TSEM is signalled once for each buffer so released.

### CSEM

A semaphore signalled by the clock ISR on every n-th activation to wake up the TIMEOUT process.

### TSEM

A semaphore signalled by the clock ISR (not, as might be expected by the transmission ISR — see Section 7.7.4) to indicate completed transmission of a buffer by the transmission ISR.

### RSEM

A semaphore signalled by the clock ISR, (not, as might be expected by the reception ISR — see Section 7.7.4) to indicate completed filling of a buffer by the reception ISR.

The procedure calls must be coordinated with waits on the semaphores. Figures 7.38(a) and 7.38(b) illustrate the code for processes IN and OUT in the case of double buffering on transmission and reception.

```
    VAR PTR : ⊦ BUFFER; I :   INTEGER;
BAD-FRAME-COUNTER := 0;
 /*GET TWO EMPTY BUFFERS INITIALLY*/
FOR I = 1 TO 2 DO BEGIN
   GET-FREE (PTR);   /*CALL TO B.P. MONITOR*/
   RECEPTION-SEND (PTR)   /*GIVE THEM TO THE ISR MODULE*/

END;

 /*GET RECEIVED FRAMES FOREVER*/
WHILE TRUE DO BEGIN
   WAIT (RSEM);   /*WAIT FOR A RECEIVED FRAME*/
   RECEPTION-GET (PTR); /*GET THE FRAME FROM THE ISR MODULE*/
   IF CHECK-CRC (PTR) AND PTR ⊦ .LENGTH >= MIN
   THEN BEGIN PUT (PTR); PTR:=GET-FREE END /*MONITOR CALL*/
   ELSE BAD-FRAME-COUNTER := BAD-FRAME-COUNTER + 1;
   RECEPTION-SEND (PTR)   /*RETURN THE BUFFER TO THE ISR*/
END;
```

*Note* — RSEM COUNTER MUST BE INITIALIZED TO 0

*Figure 7.38(a): The IN Process (Illustrating Double Buffering)*

```
    VAR I : INTEGER; PTR : ⊦ BUFFER;
 /*GET FIRST TWO FRAMES TO TRANSMIT*/
FOR I := 1 TO 2 DO BEGIN
GET (PTR);   /*DLC MONITOR CALL*/
   ADD-CRC (PTR);
   TRANSMISSION-SEND (PTR)   /*GIVE IT TO ISR*/

END;
 /*TRANSMIT FRAMES FOREVER*/
WHILE TRUE DO BEGIN
   WAIT (TSEM);   /*WAIT FOR SPACE IN THE ISR'S OUT QUEUE*/
   GET (PTR);   /*GET A BUFFER TO SEND*/
   ADD-CRC (PTR);   /*FRAME IT*/
   TRANSMISSION-SEND (PTR)   /*GIVE IT TO ISR*/

     END;
```

*NOTE* — TSEM COUNTER MUST BE INITIALIZED TO 0

*Fig 7.38(b) The OUT Process (Illustrating Double Buffering)*

### 7.7.3 Operation of the ISR Module

The ISR module consists of a set of private variables, procedures, and interrupt service routines. The private variables are accessed only by the three interrupt service routines (clock, transmit, receive) and the procedures (which are called by processes OUT and IN). Since such access is performed under conditions of interrupt lockout, it constitutes a critical section. Details of the private variables are not important here; they include pointers and counters for control of queues and character positions, as well as indicators of the state of the transmission and reception FSMs.

To cause the transmission of a buffer, the calling process in procedure TRANSMISSION$SEND queues the buffer pointer at the head of the transmit queue and, if the transmit ISR is in the idle state, the calling process makes it active and initiates the first interrupt by enabling transmission. After the calling process exits from the procedure, the transmit ISR moves one byte from the buffer in memory to the USART on every activation. When the buffer has been completely transmitted, the transmit ISR increments counter TC as notification of the fact and moves on to the next queued buffer. If there is no such buffer the ISR simply enters the idle state.

Figure 7.39 shows the relation between buffer format and transmitted frame format, that of transparent Binary Synchronous Communications (BSC). The extra characters used for framing and transparency are added by the transmit ISR in accordance with the transmission FSM shown in Figure 7.40(a). On each activation, the ISR finds itself in a particular state, performs the corresponding action, and possibly changes the state for the next activation.

Actions on the receive side are similar. The calling process places empty buffers at the head of the receive queue in procedure RECEPTION$SEND and picks up filled buffers from the appropriate slot in the same queue. The receive ISR meanwhile is filling the buffer at the tail of the queue. There is no need to activate the receive ISR, which is permanently enabled. It simply fills the current buffer from the USART one character at each activation. When done, it increments counter RC to indicate the fact, moves on to the next buffer, and places the USART in hunt mode to scan for the next SYN character. If there is no buffer available at the time the body of the next frame arrives, characters (and hence the entire frame) are lost.

By contrast with the transmit ISR, the receive ISR deletes framing and transparency characters from the incoming frame, storing only the remaining characters in the buffer, in accordance with the reception FSM of Figure 7.40(b). Every activation results in an action consistent with the current state, as well as a possible state change.

In addition to the variables mentioned above, both transmission and reception ISRs maintain a count of the number of times each FSM state is entered. During software and hardware testing this information allows a check of the framing functions of the module.

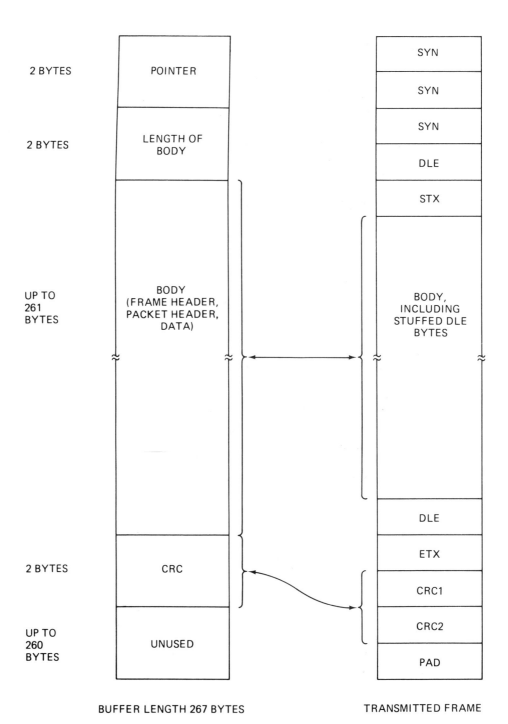

*Figure 7.39: Relation between Memory Buffers and Transmitted Frames*

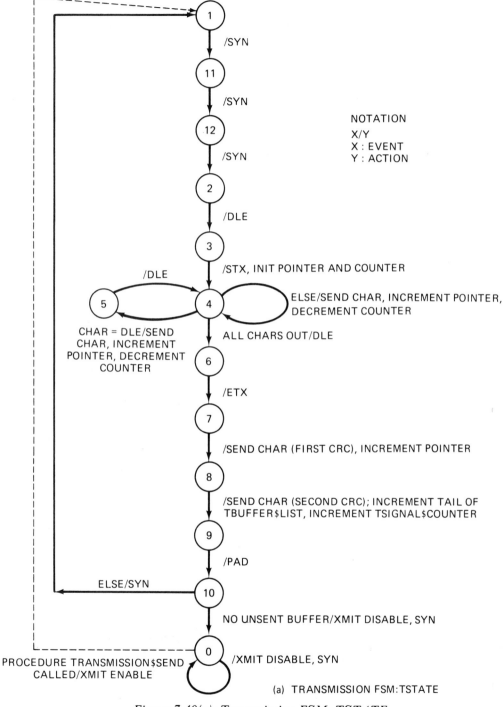

*Figure 7.40(a): Transmission FSM: TSTATE*

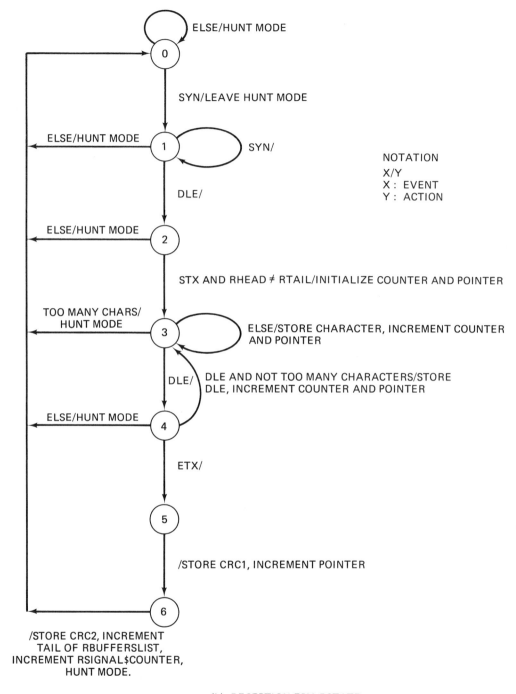

(b) RECEPTION FSM: RSTATE

*Figure 7.40(b): Reception FSM: RSTATE*

### 7.7.4   Interrupt Structure

Block mode I/O without DMA is normally set up so that each interrupt causes a datum to be moved between device and memory, and when the count has expired the interrupt service routine signals a semaphore. This simple approach cannot be adopted in general in the communications interface, however, because the character transmission time at high data rates may be less than the maximum kernel execution time. Since kernel calls are normally non-interruptable, occasional character loss would occur, during multi-buffered, synchronous I/O, leading to excessive retransmissions.

The solution is to separate the interrupt sources for the processor handling the lines into two sets, high priority and low priority, and to provide a special kernel entry for the communication processor which allows kernel execution to be interruptable. Low priority ISRs can call the kernel if they first mask off all low priority sources, and unmask them after return from the kernel. High priority sources can interrupt the kernel, but are not allowed to make any kernel calls.

The Transmission and Reception ISRs are high priority and cause movement of data between memory and USART, but make no kernel calls. When the data movement is complete, the high priority ISR initiates a low priority interrupt service request by incrementing counter RC or TC for examination by the lower priority Clock ISR.

The service routine of the clock used for process switching scans these two counters on every activation. If one is non-zero the clock ISR repetively decrements it and signals whichever semaphore RSEM or TSEM is appropriate. In this way character transfer at high data rates is not impeded by kernel execution, and interrupt sources which are allowed to call the kernel still see it as non-interruptable.

Mutual exclusion on the counters RC and TC is not a problem if the microprocessors employed have increment and decrement memory instructions which are interlocked on an instruction basis (as does the Intel 8080, for example). Then the high priority Transmission and Reception ISRs may safely interrupt the lower priority clock ISR and increment the two counters at any time in the latter's test-and-decrement-counter cycle, without fear of inconsistent results.

### 7.7.5   Synchronous versus Asynchronous ISR Modules

Experience shows that synchronous ISRs have the advantages of greater noise immunity and about 25% greater throughput due to the absence of start and stop bits. The actual software differences are relatively small if character oriented transmission is employed in both cases.

The basic difference appears in transmission format of the characters comprising a frame. Characters in asynchronous (start/stop) format are delimited by an initial mark-to-space transition for the start bit, and by 1 or 2 stop bits. Bit timing is maintained with respect to the first transition. In synchronous format, on the other hand, characters are adjacent within a single frame and there is no character start indicator. Bit timing must be supplied on an RS232C lead, usually by the data set, and character sync is obtained by a continuous scan for one or more special SYN characters preceding each frame.

The main differences in the software are in the initialization procedures and in the reception service routine. The initialization procedures must set the USART differently. And the reception ISR must place the USART in HUNT mode at the end of every frame, whether aborted or not. In asynchronous format this action is clearly inappropriate since a character sync indicator is supplied with every character.

### 7.7.6 External Specifications of the CRC Module

The CRC (cyclic redundancy check) is a two byte checksum appended to every frame before transmission. Since its value is a function of every byte in the frame, most transmission errors are detectable by a discrepancy between a CRC calculated from received data and the received CRC itself. Specifically, the CRC is the remainder when a special polynomial divides the bit sequence of the frame, treated as a polynomial. Included in the calculation are all bytes following STX and including ETX, but excluding DLE bytes stuffed for transparency. The module procedures are as follows:

        ADD$CRC (BUFF$PTR)

A procedure which, when given a buffer in the format of Figure 7.39, computes the two-byte CRC and places it at the end of the part of the buffer to be transmitted. The location of these two bytes therefore depends on the length of the frame specified in the third and fourth bytes.

        CHECK$CRC (BUFF$PTR)

A function which computes the two CRC bytes for the data of a received buffer in the format of Figure 7.32 and compares them to the two bytes already in the CRC position. If they match, the value 1 (true) is returned, otherwise the value 0 (false) is returned.

We shall not go into details of the CRC algorithm here. Suffice it to say that bit-by-bit computation is totally impractical in software because of the CPU load and that efficient, table-driven, byte-by-byte algorithms must be used.

The requirement for the CRC module disappears if a hardware CRC chip is used.

## C.   THE VIRTUAL MACHINE

The system levels comprising the virtual machine are:

— the hardware (Level 5 of Chapter 6)
— the kernel (Level 4 of Chapter 6) and
— the CHLL interface (required to convert a standard high level language into a CHLL)

Our purpose in Part C of this Chapter is to explore the relationship between the design decisions which must be made for the virtual machine and those already made at the CHLL level (Part B) for the example system. As has been pointed out previously, most of the basic design decisions are made from the "edge-in" at the CHLL and hardware levels. And, as exemplified by Part B, design decisions affecting the hardware are often made at the CHLL level. Very little design freedom is left at the kernel level except to minimize to the greatest extent possible memory requirements and execution time. As for the CHLL interface level, it is determined almost solely by the characteristics of the high level language to be used.

## 7.8   The Hardware for the Example System

This section is concerned with Level 5 of the design process described in Chapter 6.

Design decisions made in Part B have assumed a multi-master hardware configfuration with public and private memory and with private I/O devices (so that processors may communicate via common memory and be functionally dedicated to application interfaces and to communications links). A multiple master, arbitrated system bus is required so that processors may gain control of the bus to access common memory as required. And processors should be capable of waiting for the bus. Processors should have a test-and-set instruction for efficient higher-level locking of common bus resources such as the kernel; failing that, a simple hardware test-and-set flag may be provided as a device on the common bus for each resource requiring protection. The software is assumed to be resident in its entirety, so dynamic memory management is not required.

A hardware configuration which satisfies these requirements was shown in Chapter 5, Figure 5.11. It may be built, for example, using standard INTEL SBC 80/20 processor boards in a Multibus backplane, with the addition of a hardware test-and-set flag for the kernel.

Required devices on the processor boards are as follows:

— private ROM and RAM

— serial I/O (USART)

— timer (for timeouts and process switching, if required)

— priority interrupt facilities

— parallel I/O (for down-line loading)

Required and desirable features of processor chips are as follows:

*Desirable*

— test-and-set instruction

— efficient context switching

*Required*

— stack or nested workspace capability

— increment and decrement memory instructions

— ability to wait (for the bus)

— priority interrupt capability

## 7.9   The Kernel for the Example System

This section is concerned with Level 4 of the design process described in Chapter 6. The functions of this level are almost entirely determined by the decisions already made for Levels 3 and 5 in Sections 7.4 through 7.8.

The kernel must provide semaphore primitives wait and signal.

A preempt primitive for time slicing is required.

The kernel must provide primitives to support monitors which allow multiple wakeups by a single process in one call to the monitor (for example to wake up all processes in the DLC monitor by one CHANGE STATUS call). An appropriate monitor type is the gladiator described in Chapter 3. Appropriate kernel primitives are those mentioned above plus block-and-signal and unblock.

In addition, the kernel must provide a primitive to allow a process to transfer its allegiance to another processor (for example to call a link monitor resident in the private memory of a separate processor).

Design of a monolithic kernel to provide these facilities is a straightforward exercise, following Chapter 5.

## 7.10   CHLL Interface

To implement the CHLL design using a standard high level language available for the chosen hardware requires a CHLL interface layer to implement the CHLL constructs missing from the standard language. This interface layer must ensure that data areas private to each process are correctly managed, that parameters are passed safely to and from monitors, that private monitor data is really private and is maintained between monitor calls, that monitor primitives are correctly implemented using the kernel primitives, that all CHLL modules are correctly initialized and that the active modules are correctly started in the multiprocessor environment.

We shall here illustrate these aspects briefly with reference to PL/M (which would be the most appropriate language to use if INTEL hardware were chosen). Later, in Chapter 8, a more comprehensive view of CHLL's and CHLL interfaces will be presented.

### 7.10.1   Private Process Data

Correct management of private process data in PL/M requires that each process has its own stack. This can be accomplished by declaring processes as subroutines (which are never called) and then separately initializing the stack pointer for each process. The automatic stack pointer initialization and reinitialization mechanisms of PL/M are then bypassed.

### 7.10.2   Parameter Passing to Monitors

Many processes may call a monitor procedure; in a time-slicing environment there is no control over when these calls occur. They may in fact occur in the middle of the parameter passing front-end of a procedure, before the first high level statement is executed. Since it is this first high level statement which locks the monitor gate, there is nothing to stop another process from calling the same procedure before the gate is locked, thereby possibly damaging the first caller's parameters. In PL/M, such damage may be prevented by encapsulating all monitor entry procedures in a reentrant subroutine "shell" which handles parameter passing, as was described in Chapter 3.

### 7.10.3   Private Monitor Data

In PL/M there is no problem maintaining private monitor data, because modules and subroutines are allowed to possess private data. (Pascal is another story — see Chapter 8).

### 7.10.4   Monitor Scheduling Procedures

Monitor scheduling procedures in PL/M to implement ENTER, EXIT, SLEEP and AWAKEN are illustrated in Chapter 3.

### 7.10.5    CHLL Module Initialization Startup

Initialization and startup logic is always a bit messy and is here complicated by CHLL and multiprocessor issues, as follows:

— Initialization code must run in the multiProcessor environment and yet is outside of the CHLL framework which virtualizes the environment. Therefore it must control the environment directly.

— CHLL modules are initialized asynchronously on different processors; they must be prevented from calling other as yet uninitialized CHLL modules (or the kernel) until they are initialized.

— PL/M does not recognize CHLL modules, yet the addresses of the active modules are required by the initialization software so that the kernel and the private process data areas may be correctly initialized.

— The multiprocessor software must be correctly loaded (however as loading from disc is outside the scope of our example system, we shall assume that system is burned in ROM).

Dealing with these issues (excluding the last) in reverse order, the following solutions may be adopted:

— CHLL processes and ISRs are declared as public procedures (which are never called) and their addresses are obtained in the separate PL/M startup software module by applying the PL/M "dot" operator to the procedure name. This operator simply returns the address of the referenced variable. Since this operator strictly applies only to variables in PL/M, this is, strictly speaking, an illegal operation; however, it has been found to work if the procedure names are declared as variables in the startup software.

— Access to uninitialized common resources in the multiprocessor environment may be prevented by arranging that on power-up the hardware test-and-set flag protecting the kernel is in the "locked" state. Because all common resources are ultimately protected by the kernel, this effectively stops access to common resources until the flag is unlocked.

— One processor is given the job of initializing the kernel and the common resources. This processor must, after entering processes into the kernel queues in the order in which they are to begin operation, set up the first running process' stack and stack pointer and mimic a normal kernel exit, including unlocking the test-and-set flag. Time-slicing will then trigger process startup on the other processors.

## D.   ISSUES

### 7.11   Extensions for Practical Protocols

There is very little in the software architecture of the example system that needs to be changed to handle practical protocols such as X.25. Most of the changes are confined to details of the algorithms of the processes and monitors.

The main change in the processes is that processes cannot wait for acknowlegments after forwarding packets because in general multiple unacked frames are allowed.

The main additions to the DLC monitor are provisions for multi-buffering of packets, for implementing the startup protocol and for handling a greater variety of protocol messages (for example, "receiver not ready" to provide flow control, "reject" to provide fast response to errors without timing out, "disconnect", to disconnect the link, etc.).

With these changes the DLC level design described here can be used for the link level of X.25. The packet level may then be implemented as an additional monitor layer above the frame transport processes, as illustrated in Chapter 9, Section 9.4.4.2. There is little new here relating specifically to concurrent software design for multiple microprocessors and so we omit the details.

### 7.12   References

Background on protocols will be found in:

L. KLEINROCK, "Principles and Lessons in Packet Communications", Proc. IEEE, Vol. 66, No. 11, pp 1320-1330

L. POUZIN and H. ZIMMERMANN, "A Tutorial on Protocols", Proc. IEEE, Vol. 66, No. 11, pp 1371-1370

C.A. SUNSHINE, "Interprocess Communication Protocols for Computer Networks", Tech Rep 105, Dec. 75, Digital Systems Laboratory, Stanford Electronics Laboratory (available from Technical Information Service, U.S. Dept. of Commerce, Springfield, Va., 22161)

An implementation of the X.25 protocol based on the ideas presented in this book is described in:

J.K. CAVERS, "Implementation of X.25 in a Multiple Microprocessor System", Proc. Int. Conf. Comm., Toronto, June 78

Chapter **8**

# Concurrent High Level Languages

## 8.0 Introduction

Structured programming of concurrent systems designed using the principles and techniques presented in this text may be accomplished using a standard, sequential high level language. Here we show how to do it and comment on limitations and constraints of the various possible approaches. The languages considered are PL/M and PASCAL. This chapter assumes knowledge of PL/M and PASCAL and of standard software development techniques.

Concurrent programming may be accomplished using a true concurrent high level language (CHLL) or using a standard sequential high level language which has been augmented by a CHLL interface and a kernel. Until true CHLLs emerge from the experimental stage, there is little choice for most projects other than to use a standard sequential (non-concurrent) language. It is the purpose of this chapter to show how this may be accomplished.

— without modifying the standard compiler, and

— without violating good software engineering principles and practices even though the language is being used for a different purpose than originally intended.

The main standard languages considered are PL/M (in Section 8.2) and PASCAL (in Section 8.3), both of which are widely available for microprocessors. We refer to the augmented languages as "Multi-PL/M" and "Multi-PASCAL", respectively. In Section 8.4 we consider broader issues associated with these approaches and with the use of CHLLs in general.

## 8.1  General Organization of a Concurrent System

A concurrent system in general will consist of the following main functional components:

a)  The Kernel and CHLL Interface

The kernel and CHLL interface includes, in addition to the kernel primitives, the following:

- the access layer, a set of procedures which makes kernel primitives easily accessible to programs written in the High Level Language;

- the CHLL interface layer, consisting of a set of scheduling procedures for synchronizing processes in monitors using kernel primitives (this layer may be shared among monitors or copies of the procedures may belong to each monitor);

- utility procedures for initializing processes, semaphores, interrupt vectors, etc.

b)  Processes

Every process is written as a separate procedure having no parameters which is never called and from which there is no return.

c)  Shared Procedures

Procedures which can be shared among processes or which regulate access to shared data structures will be compiled separately according to the conventions of the language. These include monitor entry procedures and reentrant utility procedures.

d)  Global Data

All loose data common to multiple processes is declared in the global data area. Depending on the language and the design philosophy adopted, this data may include shared variables such as semaphores and monitor variables, as well as the PCB and stack for every process in the system.

e)  ISRs (Interrupt Service Routines)

Interrupt service routines can usually be coded in Assembly Language or in the High Level Language. In either case they are separately compiled.
If the ISR is coded in a High Level Language, it is declared as a parameterless procedure.

If time slicing is desired a clock ISR must be provided. A typical clock ISR makes periodic calls to the preempt primitive on behalf of the interrupted process. More elaborate clock ISRs might include signals against global semaphores to provide process delay and timeout services.

### f) Initialization Code (The "Main Program")

The programmer must provide a main program for each processor. In aggregate, this code simply initializes all shared variables and interrupt vectors, etc., then disappears into the kernel to let the multiprogramming and multiprocessing action begin.

The functions to be performed are as follows. Multiprocessing considerations were discussed in Chapter 7 (Section 7.10). First the kernel is initialized. The remaining initialization calls can be performed in any order. Semaphore initialization must be performed if semaphores are not fixed by kernel initialization. Processes must be declared by a call which binds the PCB and stack area of a process, initializes them and puts the PCB on a ready to run queue. Interrupt vectors must be set to point to the appropriate ISRs. Next, all monitors and other shared structures can be initialized by calls to the appropriate CHLL procedures. Finally, the main programs call procedures from which they never return. Within these procedures the clock interrupts are enabled and the first ready to run processes are dispatched. At this point the concurrent system is off and running.

## 8.2   Multi-PL/M

It is the purpose of this section to provide guidelines for the use of PL/M as a concurrent high level language. Knowledge of the use of PL/M as a sequential programming language and of standard software development practices is assumed. Some material of this section has been covered previously in Chapter 7 (Section 7.10). No compiler modifications or illegal programming tricks are required.

### 8.2.1   Defining Concurrent Components in PL/M

Processes and monitors are not recognized by the PL/M compiler. To implement them successfully they must be defined in a particular way in PL/M, with naming conventions and comments to indicate their nature. The form of definition is illustrated in Figures 8.1 and 8.2. Note that processes are defined as procedures and monitors are defined as modules. Before proceeding further some properties of PL/M need explanation.

```
PROCESSES IN PL/M

  /*PROCESS MODULE*/
  CONTROL$PROCESS$MODULE DO;

    DECLARE ....     /*PROCESS VARIABLES*/
    ...

      CONTROL$PROCESS:PROCEDURE PUBLIC; /*PROCESS CODE*/
        DO FOREVER;
        ...
        END;
      END CONTROL$PROCESS;

  END CONTROL$PROCESS$MODULE;
```

*Figure 8.1: Processes in PL/M*

```
MONITORS IN PL/M

  /*MONITOR MODULE*/
  PACKET$MON: DO;
    ENTER:PROCEDURE; /*PRIMITIVES*/
          CALL KERNEL...
          ...
    EXIT: ...
    SLEEP:...
    ...
    DECLARE ...; /*SHARED VARIABLES*/

      /*ENTRY PROCEDURES*/
      PLACE$CALL$PACKET$MON: PROCEDURE ... REENTRANT PUBLIC;
                             DECLARE ...; /*PRIVATE VARS*/
                             CALL ENTER;
                             ...
      END PLACE$CALL$PACKET$MON;
      WAIT$FOR$CALL$PACKET$MON: ...
                             ...

  END PACKET$MON;
```

*Figure 8.2: Monitors in PL/M*

```
ENTER;
    PROCEDURE;
        CALL KERNEL (0312H); /*WAIT (GATE)*/
    END ENTER;
    DEPART;
    PROCEDURE;
        IF COUNTER > 0
            THEN CALL KERNEL (0413H,0); /
*SIGNAL (HIGH PRIORITY)*/                    .
            ELSE CALL KERNEL (0412H,0); /*SIGNAL (GATE) */
        END DEPART;

    SLEEP;
    PROCEDURE (ADDRESS$OF$QUEUE);
        DECLARE ADDRESS$OF$QUEUE ADDRESS;
        DECLARE QUEUE BASED ADDRESS$OF$QUEUE BYTE;
        QUEUE= NAME
        IF COUNTER > 0
            THEN CALL KERNEL (0513H,0); /*BLOCK AND SIGNAL
                                              (HIGH PRIORITY)*/
            ELSE CALL KERNEL (0512H,0); /
*BLOCK AND SIGNAL (GATE)*/
        CALL KERNEL (0313H,0); /*WAIT (HIGH PRIORITY)*/
        COUNTER=COUNTER-1;
    END SLEEP;

    AWAKEN:

    PROCEDURE (ADDRESS$OF$QUEUE)
        DECLARE ADDRESS$OF$QUEUE ADDRESS;
        DECLARE QUEUE BASED ADDRESS$OF$QUEUE BYTE;
        DECLARE TEMPORARY BYTE;
        TEMPORARY = QUEUE;
        IF QUEUE  <> NIL THEN
            DO;COUNTER = COUNTER + 1;
                QUEUE = NIL;
                CALL KERNEL (0600H + TEMPORARY, 0); /
*UNBLOCK(TEMPORARY)*/
            END;
    END AWAKEN;
```

*Figure 8.3: Example of Monitor Scheduling Procedures in PL/M (Monitor of Type Gladiator)*

## 8.2.2  About PL/M

Placing code at the module level in PL/M, as opposed to the procedure level, causes the compiler to generate object code that manipulates the stack pointer directly. For stand alone modules that run in a non-multiprocessing environment this is not a problem and in fact could be useful. For modules that run in a multiprocessing environment this can create problems.

Code at the module level results in the compiler generating object code which manipulates the stack as follows:

1) Code which initializes the stack pointer is placed before the code of the first executable statement of the module. The stack is initialized to the value specified at locate time by the LOCATE control keyword "STACK(address)".

2) The stack is re-initialized before every procedure call.

Such stack manipulations are clearly undesirable for processes in a multi-processing system as each process must have its own independent stack area. In order to avoid any potential problems all process code should be contained within the scope of a procedure. This will ensure that the stack pointer is always manipulated in relative terms and not in absolute terms. Processes will be declared as procedures which are never called and which will contain an infinite loop so that a return will never be made. One or many processes may be defined within a single PL/M module.

The setup of all process stack pointers is the responsibility of the startup software and the stack pointer should be manipulated in absolute terms only by the kernel.

Declaring monitors as modules causes no problems because the only executable code of the monitor is in its procedures and these are always called from outside the monitor module. And the module construct allows each monitor to have its own data, hidden from other system modules but commonly accessible to all procedures of the monitor. Furthermore, the division of the software system into modules in this way is an aid to orderly software development; it is good software engineering practice.

### 8.2.3  Implementation Techniques

#### *8.2.3.1  Monitor Parameters*

As described in Chapter 3, parameters must be passed to PL/M monitors reentrantly. This avoids parameter damage due to process switching before the monitor gate is locked by the first high level language statement of the entry procedure. This may be accomplished by declaring all monitor entry procedures which have parameters as reentrant in their entirety, which may be inefficient. Or it may be accomplished by separating the body of each such procedure and declaring it as a separate internal monitor procedure, leaving only a reentrant "shell" as the entry procedure.

### 8.2.3.2 Monitor Scheduling Procedures

Monitor scheduling procedures (ENTER, EXIT, SLEEP, AWAKEN, etc.) may be defined as private procedures of each monitor module. This approach results in simple code for the primitives because the gate semaphores and condition variables involved are known to the procedures and do not have to be specified by the caller. This approach was used in the examples of Chapter 3 and is illustrated in Figure 8.3 using PL/M. Assumed in Figure 8.3 is a kernel of the type given as an example in Chapter 5.

Alternatively, monitor scheduling procedures they may be defined as shared procedures of a CHLL interface layer which is used by all monitors. This approach results in slightly more complex code for the primitives, but is more flexible and less wasteful of memory. This approach is illustrated in the next section using PASCAL. Of course either approach may be used in either language.

### 8.2.3.3 Naming Conventions

Clearly Multi-PL/M uses PL/M procedures for many different purposes. In order to facilitate the functional type identification of procedures, keywords may be added as suffixes to their names ("name" identifies the chosen name while the capital letters compose the keyword):

name$INIT: (initialize); identifies any procedure that initializes data structures, devices, etc., and is called by the startup software.

name$process: identifies any procedure that contains a process.

name$ISR: (interrupt service routine); identifies any procedure that is of type interrupt.

name$MON: (monitor); identifies any entry procedure in a monitor.

### 8.2.3.4 Initialization

Difficulties may be encountered while trying to determine the starting address of processes and ISRs in the startup module software. Process starting addresses are needed by the kernel when processes are declared. The starting addresses for the ISRs are needed for setting up the interrupt jump table.

Addresses can be obtained manually. The symbol table listing, generated by the locator, showing the physical address of all symbols, can be used to locate each symbol. Each address can then in turn be entered by hand into the startup software module which then can be compiled, linked and located and ultimately down line loaded with the other modules. The manual intervention is, however, an unnecessary evil.

The following is a fully automated method for obtaining addresses:

1) Processes and ISRs are declared as procedures, thus each has a unique symbol identifier and is assigned with PUBLIC attribute

2) Any symbols whose locations need be known by the startup software are declared as external byte variables within the startup software module

3) A reference to the address of the externally defined variables results in the required address (i.e., use of the dot (.) operator)

Obtaining the addresses of labels in this fashion, although not strictly legal in PL/M, is possible because the linker only flags the discrepancy of the label types thus created and continues to make the link anyway. The programmer is notified of the discrepancy by the message "label, has differing types" appearing in the listing generated by the linker; this message may be ignored.

### 8.2.3.5  *Developing and Running a Multi- PL/M System*

To develop and run a concurrent system using PL/M requires the following steps (ignoring testing and debugging for the moment):

1) The kernel is coded (usually in assembler, although those parts of it which do not manipulate the machine registers directly could be written in PL/M). Any CHLL interface routines required to implement shared monitor primitives are coded.

2) The processes and monitors are written in PL/M following the guidelines presented in this chapter.

3) The startup code is written in PL/M following the guidelines presented in this chapter and in Chapter 7 (Section 7.10).

4) PL/M source modules with their associated compiled object modules are stored on a number of diskettes as required. Each time a source module is updated it is immediately compiled so that object modules accurately represent associated source modules.

5) All required object modules for this particular system to be run are transferred to a Load Module Generation (LMG) diskette.

6) All the object modules on the LMG diskette, as well as the PL/M library routines on the PL/M diskette, are linked together to form one large object file. Note that not all the modules need be combined to form the object file. Some modules, such as the kernel entry routines, the kernel and debugging routines, which are already known to be bug-free are simply linked without combining. The only routines that need be linked and combined are those that are currently under development.

7) The relocatable object module created above is located in the required area of memory. The resulting module is the load module and will be down line loaded. When locating a module it will likely be desirable to generate a diagnostic

listing which contains all the labels and their corresponding addresses. With experience we have found that having the locator include the starting address of each line of PLM code in the symbol table listing is of minimal usefulness. With this information included we found that the locator took longer to do its job as the symbol table was longer. Knowledge of where each line of source code is located is only of value if the programmer wants to:

— verify correct compiler operation (should not be necessary),

— insert physical break points (not very useful — see Chapter 9), or

— fix a bug at the machine code level (not recommended).

8) The load module is down line loaded. Often several load modules have to be loaded, the others being modules of system routines and utilities that had previously been located.

A number of steps in the above procedure can be somewhat automated. Steps 6 and 7 are most prone to error as the command string used to invoke the linker and locator is rather involved. There may be many modules that must be linked, some with and some without combining. The locator requires that the module starting addresses for code and data be specified among the appropriate diagnostic switches. Overcoming the problem of entering the command strings each time a new load module is generated is a simple matter of using the development system SUBMIT facility. The SUBMIT command causes command sequences to be taken from a file on diskette rather than the console. The file can be created using the text editor. Once created the submit file may be submitted for batch processing as often as necessary.

## 8.3  Multi-Pascal

Pascal may be used as a concurrent programming language in much the same way as PL/M. However, differences in the language structure make certain of the guidelines different. It is the purpose of this section to provide appropriate guidelines for Pascal. Knowledge of the use of standard Pascal as a sequential programming language and of standard software development practices is assumed. As with PL/M, no compiler modifications or illegal programming tricks are required.

### 8.3.1  Defining Concurrent Components in Pascal

Processes and monitors are defined in Pascal as illustrated in Figures 8.4 and 8.5. Note that as with PL/M processes are defined as procedures. Monitors are defined slightly differently because of Pascal's lack of permanent local variables. The type definitions to support processes and monitors as illustrated are described in Section 8.3.3.

```
/* IDLE PROCESS VARIABLE DECLARATIONS */
    CONST SIZEIDLE=10;
    VAR     PCBIDLE:PCB;
            STACKIDLE:RECORD LOWAREA:ARRAY[1..SIZEIDLE]
                      OF INTEGER; HIGHAREA:STACK;
                      END;

    /*IDLE PROCESS CODE*/
    PROCEDURE IDLE;
    BEGIN
        WHILE TRUE DO PREEMPT
    END;
```

*Figure 8.4: An Example of a Process in Pascal*

### 8.3.2  About Pascal

There are a number of language and implementation features of standard Pascal which must be noted in order to use the language effectively (and to stay out of trouble).

1. All procedures are reentrant, since arguments and local variables are maintained in the stack of the calling process. This is a boon to writing system programs which make extensive use of shared procedures. There is an ugly flip side of the coin, however: first, all processes calling the procedure must have adequate stack space; and second, the local variables evaporate upon return from the procedure. The latter property means that all variables to be retained from call to call or shared between processes must be declared as global level variables. They are thereby made accessible to any other process, which defeats attempts at the information hiding in monitors important to survival in a large concurrent environment.

2. Separate compilation implies that any global variables declared in a file must be compiled with the declarations of all other global variables or they will overlap each other when linked. To avoid this problem, it is necessary to create a single file containing all global variables and compile each file with it.

3. As in PL/M, the name of a Pascal procedure is a symbol which has for its value the address of its first executable statement (used in initialization). However, note that the first executable statement of a Pascal procedure is not the first statement of the source code; it is instead a resetting of the stack pointer to make space for the local variables.

```
/*MONITOR DECLARATIONS FOR GLOBAL DATA AREA */
    CONST BBSIZE=16; BBEND=15;
    VAR BBUF: RECORD
            MON:CTRL;
            NONFULL:EVENT;
            NONEMPTY:EVENT;
            HEAD:0..BBEND;
            COUNT:0..BBSIZE;
            STORE:ARRAY[0..BBEND] OF INTEGER;
            END;
/* MONITOR CODE */

    PROCEDURE BBPUT(X:INTEGER);
        BEGIN
        WITH BBUF DO BEGIN
            ENTER(MON);
            IF COUNT=BBSIZE THEN STALL(NONFULL);
            STORE[(HEAD+COUNT) MOD BBSIZE]:=X;
            COUNT:=COUNT+1;
            PROCEED(NONEMPTY);
            LEAVE(MON)
            END
        END;

    FUNCTION BBGET:INTEGER;
        BEGIN
        WITH BBUF DO BEGIN
            ENTER(MON);
            IF COUNT=0 THEN STALL(NONEMPTY);
            BBGET:=STORE[HEAD];
            HEAD:=(HEAD+1) MOD BBSIZE;
            COUNT:=COUNT-1;
            PROCEED(NONFULL);
            LEAVE(MON)
            END
        END;

    PROCEDURE BBINIT;
        BEGIN
        WITH BBUF DO BEGIN
            HEAD:=0;
            COUNT:=0;
            INITCTRL(MON);
            INITEVENT(NONFULL,MON);
            INITEVENT(NONEMPTY,MON)
            END
        END;
```

*Figure 8.5: Monitors in Pascal*

### 8.3.3  Implementation Techniques

Many of the implementation techniques carry over from those recommended for PL/M. And the consequences of major language differences have already been noted. However PASCAL's powerful type definition facility can provide significant advantages. In the examples of Chapter 3 each monitor was assumed to have its own copy of the scheduling procedures. And each of these procedures was assumed to know the names of all control semaphores and condition variables for that monitor (e.g., the gate and pending variables for the monitor of type Mediator and the gate and eligible variables for the monitor of type Gladiator — Figures 3.12 and 3.14). To make these procedures shareable, these variables must be passed as parameters along with the condition variable involved. Although this may be done in PL/M, it has not been illustrated. An elegant way of doing this in Pascal to provide support for both Mediator and Gladiator monitors is illustrated in Figures 8.6 and 8.7.

```
TYPE SEMAPHORE=RECORD HEAD,TAIL: ↑ PCB;
                      COUNT:INTEGER
          END;
     TSKPRIO=1..2;
     CPUPRIO=0..7;
     SEMPTR=↑SEMAPHORE;;
     PCB=RECORD PLINK: ↑ PCB;
            PPRIOR:TSKPRIO;
            PSTACK:↑STACK;
            PSTATE:INTEGER;
            PSEM:SEMAPHORE
          END;

     STACK = RECORD STARTREG:ARRAY[0..5] OF INTEGER;
            PC,PSW:INTEGER
          END;

     CTRL=RECORD GATE:SEMAPHORE;
            HIPRIWAITING:INTEGER;
            HIPRI:SEMAPHORE;
          END;

     EVENT=RECORD CTRLLINK: ↑ CTRL;
                  EVWAITING:INTEGER;
                  EVSEM:SEMAPHORE;
          END;
```

*Figure 8.6: Type Definition to Support and Shared Monitor Scheduling Procedures*

```
PROCEDURE ENTER (VAR C:CTRL);
   BEGIN
         WAIT(C.GATE); END;

PROCEDURE LEAVE(VAR C:CTRL);
   BEGIN
         WITH C DO
               IF HIPRIWAITING > 0 THEN BEGIN
                       HIPRIWAITING:=HIPRIWAITING-1;
                       SIGNAL(HIPRI);
                       END
               ELSE SIGNAL (GATE) END;
PROCEDURE SLEEP(VAR E:EVENT);
   BEGIN
         WITH E DO BEGIN
               EVWAITING:=EVWAITING+1;
               LEAVE(CTRLLINK ↑);
               WAIT(EVSEM);
               WAIT(CTRLLINK ↑.HIPRI); END;

   END;
PROCEDURE AWAKEN(VAR E:EVENT);
   BEGIN
         WITH E DO BEGIN
               IF EVWAITING > 0 THEN BEGIN
                 EVWAITING:=EVWAITING-1;
                 CTRLLINK↑.HIPRIWAITING:=CTRLLINK↑
                 .HIPRIWAITING+1; SIGNAL(EVSEM); END

         END;
   END;
PROCEDURE STALL(VAR E:EVENT);
    BEGIN
         WITH E DO BEGIN
            EVWAITING:=EVWAITING+1;
            LEAVE(CTRLLINK ↑);
            WAIT(EVSEM) END
   END;
PROCEDURE PROCEED( VAR E:EVENT);
   BEGIN
         WITH E DO BEGIN
            IF EVWAITING > 0 THEN BEGIN
               EVWAITING:=EVWAITING-1;
               CTRLLINK ↑ .HIPRIWAITING:=CTRLLINK
               .HIPRIWAITING+1; SIGNAL(EVSEM);
               WAIT(CTRLLINK↑.HIPRI)
               END
         END
   END;
```

*Figure 8.7: Shared Monitor Scheduling Procedures*

## 8.4   Language Issues

The techniques and guidelines described in Section 8.2 and 8.3 may of course be applied using variants of PL/M and Pascal, with due care for differences in the compiler and development software. They may even be applied to such "unfriendly" languages as Fortran, although with some awkwardness because most Fortrans do not support reentrancy. In the authors' Laboratory they have been applied using INTEL's PL/M and using the PASCAL compiler for the PDP11 line of computers distributed commercially by Oregon Minicomputer Software Inc. But what of true concurrent High Level Languages?

A number of experimental CHLLs have been developed in research institutions and universities. Perhaps the most widely known and distributed is Brinch Hansen's Concurrent Pascal. Others include Wirth's Modula and Holt's CSP/K. All of these are PDP-11 and uniprocessor oriented. One laboratory is known by the authors to have developed and used in distributed data base applications a true CHLL. Others are experimenting with the development of such languages.

The only manufacturer known to the authors to have developed a true CHLL for specific use with microprocessors is Texas Instruments. TI's Microprocessor Pascal is said to support processes, semaphores and shared files. However, it appears not to support multiple processors. And of course it is only available for TI processors.

Intel has developed a kernel called RMX80 which supports processes and interprocess communications. The kernel may be used by assembly language or PL/M programs. But it is not by itself a CHLL and it does not support multiple processors.

In general, then, until CHLL's come of age, multiple microprocessor system designers and programmers are on their own. What is lost by not using a true CHLL? Experience in our Laboratory and discussions with others working in the same area suggest that the major loss is of automatic rechecking that the declared access graph is still being obeyed when interfaces in a system under development are changed. This loss can be counterbalanced by greater administrative control of changes to systems under development.

## 8.5   References

Language descriptions for standard PL/M and Pascal are provided in:

"PL/M-80 Programming Manual", Intel Corporation

KATHLEEN JENSEN and NICKLAUS WIRTH, "Pascal User Manual and Report", Second Edition Springer-Verlag, 1975 Concurrent high level languages are described in:

PER BRINCH HANSEN, "The Programming Language Concurrent Pascal", IEEE Trans. Software Engineering, Vol. SE-1, No. 2, June 75

N. WIRTH, "MODULA: A Language for Modular Multiprogramming", Report 18, Eidgenössische Technische Hochschule Zurich, Institut für Informatik, Zurich, Switzerland

R.C. Holt, G.S. Graham, E.A. Lazowska, M.A. Scott, "Structured Concurrent Programming with Operating Systems Applications", Addison-Wesley, 1978

Roger R. Bate and Douglas S. Johnson, "Pascal Software Supports Real-Time Multiprogramming on Small Systems", Electronics, June 7, 1979

The RMX80 kernel is described briefly in:

Kenneth Burgett and Edward F. O'Neil, "An Integral Real-Time Executive for Microcomputers", Computer Design, July 1977

# ISSUES IN DESIGN AND DEVELOPMENT

In the exponential growth environment which we all function capabilities and mechanisms change rapidly. However, the issues which must be resolved tend to remain, and to demand responses.

In this part on-going issues will be discussed and alternatives proposed and evaluated. Certainly such things as reliability, efficiency, hardware transparency, the distribution of control and other factors will always be faced by designers regardless of the mechanisms for implementation.

Chapter **9**

# Systems Issues

## 9.0  Introduction

In this chapter we return to some key issues which are important to the success of advanced design and development work in this area, but which could only be treated briefly in previous chapters.

## 9.1  Key Issues

The first issue to be considered, in Section 9.2, is that of transparency of the hardware architecture. When the number of processors and/or the nature of the interprocessor interfaces are changed, what are the effects on software? As was shown at considerable length in Chapter 7, the software designer using monitors and monitor-like structures must take the ultimate hardware architecture into account. Can software be designed using other techniques in such a way that such changes are transparent? What are the limitations on transparency?

Another key issue is reliability and, in particular, recovery from failure. It was shown in Chapter 7 how software protocols may be designed to protect nodes of a network from failure in other nodes and in the inter-node links. In general such techniques may be employed between system modules to limit the effects of failures in individual modules or in the inter-module interfaces. The subject is closely related to that of transparency and is also treated in Section 9.2.

Another key issue to be considered, in Section 9.3, is efficiency. The potential efficiency problems with a monolithic software kernel were discussed in Chapter 5. Section 9.3 considers software and hardware approaches to making the kernel more efficient.

The final key issue to be considered, in Section 9.4, is system testing and debugging, with particular emphasis on those aspects that are unique to multiple microprocessor systems implemented using the CHLL techniques of this book.

245

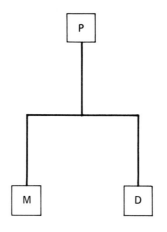

(a) BEFORE

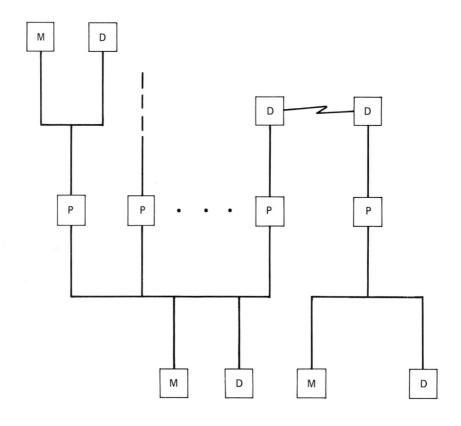

(b) AFTER

*Figure 9.1: Processor Transparency*

Two important issues are not treated here, both for lack of space and because they do not apply only to multiple processor systems. The first is project organization and control. And the second is the provision of good development facilities for both hardware and software engineers. Both issues interact to impact cost. And current solutions are far from satisfactory.

The main sections of this chapter may be read independently in any order, depending on the reader's interest.

## 9.2    Transparency

The term transparency here refers to the situation when software does not have to be recompiled when changes are made in the following items:.

1.  Type and number of processors *(processor- transparency)* — Figure 9.1 shows a possible change from one to many processors which may be required to be transparent.

2.  Type and bandwidth of interprocessor interfaces *(interface-transparency)* — Figure 9.2 shows a change from a high bandwidth, memory-coupled interface to a low bandwidth, thin-wire coupled interface which may be required to be transparent.

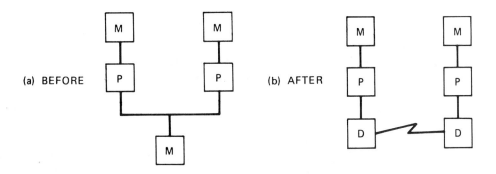

*Figure 9.2: Interface Transparency*

Note the implications of processor transparency. Programs and data may reside after a change in private or common memory of a "memory-coupled" multiprocessor or in the private memory of a separate "thin-wire coupled" processor. Under what conditions is it possible and desirable to do this without recompilation?

Note the implications of interface transparency. High bandwidth interfaces are implemented, typically, by hardware techniques which are as reliable as the processors themselves. However, low bandwidth interfaces are often implemented by hardware techniques which are orders of magnitude less reliable than the processors. High bandwidth interfaces may be assumed to be error free in the sense that data and instructions sent across the interface are no more likely to fail than are the processors on each side of the interface. On the other hand, low bandwidth interfaces must be prepared to cope with transmission errors. Considering these differences, is full interface transparency possible or desirable?

The terms "memory-coupled" and "thin-wire coupled" are used here to characterize both hardware and software techniques for coupling processors. Applied to hardware their meaning is direct and obvious. Applied to software they distinguish interprocess communications techniques according to whether or not they require common memory.

This section considers transparency issues under three headings:

— Software Structures (Section 9.2.1);

— Memory-Coupled Multiprocessor Architectures (Section 9.2.2);

— Thin Wire Coupled Architectures and Reliability (Section 9.2.3).

Conclusions are reached in Section 9.2.4.

## 9.2.1   Software Structures

Software structures may affect transparency because of the type of interprocess communication involved. Interprocess communication techniques are categorized below as "memory- coupled" or "thin-wire coupled". Then the two major approaches of Monitors and Message Passing are discussed, with special emphasis on those features which may affect transparency.

(a) *Interprocess Communication Techniques*

One method of coordinating processes uses shared data and programs called MONITORS (we include synchronizing tools such as semaphores in this category).

In another technique, processes coordinate by exchanging MESSAGES. If processors are separated by large distances a COMMUNICATION SUBNETWORK must also exist.

Monitors and monitor-like structures may be characterized as memory-coupled because shared memory is required. Memory-coupled software may be made processor-transparent but is not easy to make interface-transparent. Message passing structures may be characterized as thin-wire coupled, because communication lines (thin wires) are sufficient for their implementation; memory coupling is not necessary. Thin-wire coupled software may in principle be made fully transparent. The question is, should full transparency be aimed at in practice by the general adoption of thin-wire coupled software interfaces?

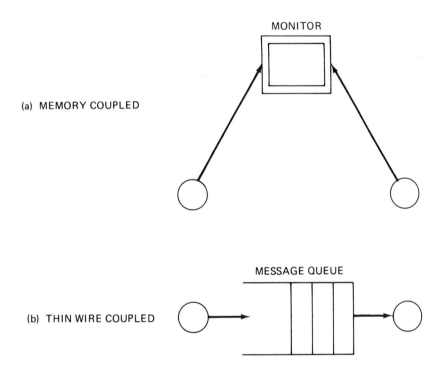

*Figure 9.3: Software Techniques*

Figure 9.3 shows how the two software techniques may be depicted graphically; Figure 9.3a shows two processes calling a monitor; Figure 9.3b shows two processes communicating via a message queue (which may be extended over thin wires).

We note in passing that monitor-based software may be transformed into message-based software by reprogramming. In particular, a monitor may be replaced by a process which waits on messages requesting execution of the entry procedure functions and which sends answers when the requested functions are completed. A process calling the monitor simply sends a request and then waits for an answer. The entry procedures are completely under the control of a single process, namely the monitor-process. There is no internal scheduling with its resulting complications. However the price of this simplicity is a more complex, lower-level, inter-process communication mechanism which must be concerned with message buffer allocation, message acknowledgement and message flow control. Whether or not the end result is worth it in a strictly memory-coupled system is an open question.

### (b) *Monitors and Transparency*

When processes can access common memory, monitors provide a powerful and efficient structure through which to coordinate their actions. At the network level, problems with the implementation of critical regions, multiple copies of data, and recovery from failures and communication problems make monitors a less desirable structure for the coordination of processes.

Chapter 7 illustrated some transparency issues which arise with monitors. Transparency is achievable when the range of hardware architectures within the memory-coupled class is known.

### (c) *Message Passing and Transparency*

When processes are loosely coupled, the exchange of messages provides an appropriate method for their coordination. The designer may want processes to be so coupled for a number of reasons. The processes may exist in different geographical locations and therefore cannot access common memory. Even if the processes exist in a memory-coupled system, the designer may want the option to separate them geographically in the future. It may also happen that the processes exist on different pages of large memory and are not capable of calling shared subroutines or accessing shared data.

The requirement for communications protocols is the basic new feature of message passing systems. Several protocol levels may be required, characterized by the several types of message acknowledgements which can be implemented, as illustrated by Figure 9.4. A message could be acknowledged after it has been put into a buffer at the receiving end. This verifies the communication link and buffer availability (LINK protocol). Or a message could be acknowledged after the buffer has been picked up by the receiving process. This verifies that processing of the message has started (MESSAGE protocol). Or, an acknowledgement could be generated after the receiving process has done something with the message (TRANSACTION protocol).

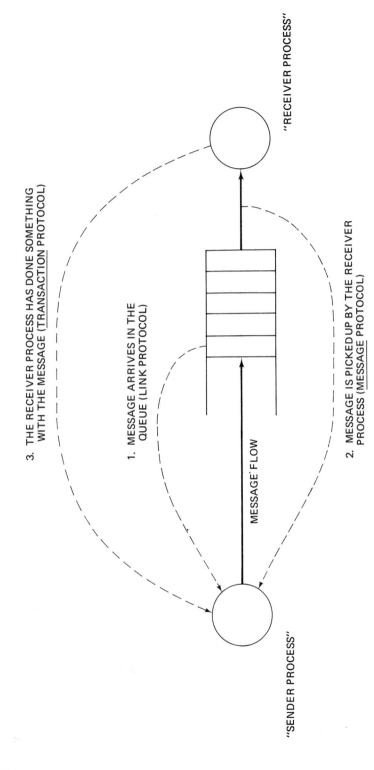

3. THE RECEIVER PROCESS HAS DONE SOMETHING WITH THE MESSAGE (TRANSACTION PROTOCOL)

1. MESSAGE ARRIVES IN THE QUEUE (LINK PROTOCOL)

2. MESSAGE IS PICKED UP BY THE RECEIVER PROCESS (MESSAGE PROTOCOL)

MESSAGE FLOW

"RECEIVER PROCESS"

"SENDER PROCESS"

*Figure 9.4: Possible Message Acknowledgements*

In these protocols, a process could pick up acknowledgements by (i) waiting immediately after sending a message, or (ii) picking them up from a special message acknowledgement queue, or (iii) receiving special messages in a general purpose input queue for the process, or (iv) receiving them as part of arriving messages.

In addition, a message passing system must be concerned with message buffer allocation, to avoid buffer depletion by a talkative process. And it must be concerned with message flow control, to enable a popular process to warn its admirers that it can receive no more messages for the present.

The issues are complicated and it is therefore difficult to design a message passing service which is both efficient and interface-transparent. Experience in the authors' laboratory with mixed hardware configurations in which both shared memory and thin wires were used for inter-processor communication suggests that a corresponding mixed software approach is both logically coherent and efficient. Monitors are used for parts of the system which have shared memory. And message passing is used between parts connected by thin wires. This is the approach presented in this book, particularly in Chapter 7. On a system-wide basis some transparency is lost. This important subject is pursued further in Section 9.2.3 where it is proposed that message-based techniques are most suitable between components of a system when communications between the components is unreliable.

### 9.2.2   Memory-Coupled Multiprocessor Architectures

Memory coupled systems may be classified as:

— uniform

— semi-uniform

— heterogeneous

Memory-coupled systems are uniform (in hardware) if:

1) all processor modules are identical

2) all memory modules are identical

3) all processors have equal access to all memory

4) all processors have equal access to all peripherals

5) all processor modules are treated as "resources" by a single central executive which allocates tasks to processors on a dynamic basis.

Figure 9.5 is a logical view of the only architecture which combines the above mentioned properties. For this architecture the multiplicity of processors is transparent

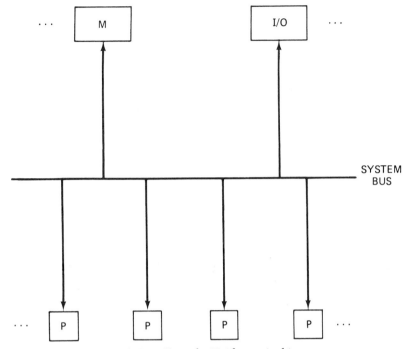

*Figure 9.5: A Strictly Uniform Architecture*

to everything above the kernel. Software development may thus proceed as if dealing with a single processor. (In fact, the same software can run on any system containing from 1 to n processors with no differences other than in system performance).

For transparency (and other) reasons no other architectures would even be considered were it not for the problem of bus congestion. When the system requires more than 2 or 3 processors, system buses, memory modules and peripherals will become system bottlenecks. Compromises on uniformity are then required, producing semi-uniform systems. Such systems should have a minimal set of necessary deviations from uniformity just sufficient to eliminate the encountered problems (introduction of private memory or I/O for example does not result in a loss of all advantages associated with uniform systems).

For tightly coupled, physically undispersed multiprocessor systems, necessary requirements and desirable features point to semi-uniform systems with interprocessor communications via shared memory. However, there are at least two main issues associated with such systems:

1) COMMON/PRIVATE memory tradeoffs must be considered. Memory allocation problems as well as bus congestion problems must be resolved with flexibility.

2) COMMON/PRIVATE I/O access tradeoffs must be resolved with flexibility.

Let us deal with the memory problems associated with semi-uniform systems. Typical multiprocessor systems may have the following unfortunate limitations:

1) If three or four processors are connected to the common system bus, shared memory quickly becomes an important system bottleneck when it is used to store frequently executed programs.

2) The boundary between private and public memory may not be sufficiently flexible. There may not be enough private memory on some boards and too much on others.

3) There may not be enough TOTAL memory (particularly if the system is based on a non-extendable 16 bit address bus).

4) Private memory may be really PRIVATE and under no circumstances may ever be reached by any other processor. This, combined with insufficient total memory, can result in severe memory allocation problems which in turn may necessitate design compromises.

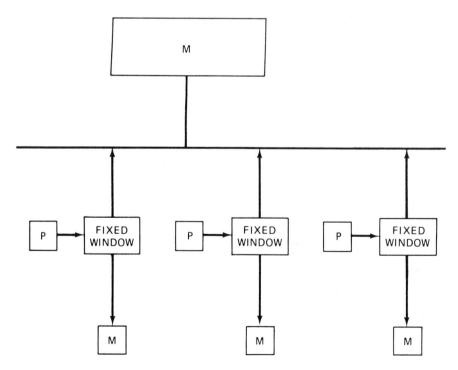

*Figure 9.6: Single Bus, Common and Private Memory with Fixed Boundary*

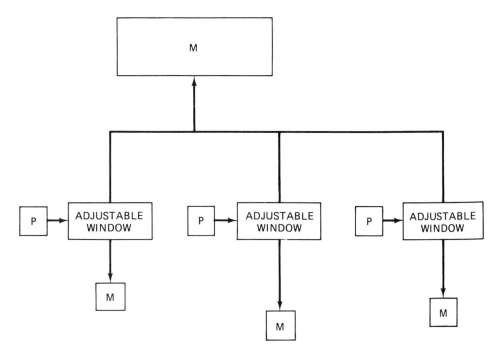

*Figure 9.7: Flexible Common/Private Memory Boundaries*

All newer processor chips can directly address megabytes of primary memory, and many offer memory management chips as a bonus. New processors seem an efficient and reasonable solution to the total memory problem.

The PRIVATE memory and BUS CONGESTION problems can be solved in several ways within a semi-uniform architecture. We will discuss three.

The first way of assigning private and public memory in a semi-uniform architecture is shown in Figure 9.6. Each processor has a fixed address range which corresponds to strictly private memory, and the remainder of the range corresponds to strictly public memory. These fixed address ranges are identical for all processors. The main problems are likely to be the limited total address space and the rigid private memory organization forced by a standard board configuration. There are, however, other problems associated with such an architecture:

1) For applications making frequent use of common memory, the common system bus may become a system bottleneck and thereby degrade overall system performance.

2) Private memory is strictly private and may under no circumstances ever be accessed by another processor. This invariably results in a loss of transparency, as was illustrated in Chapter 7.

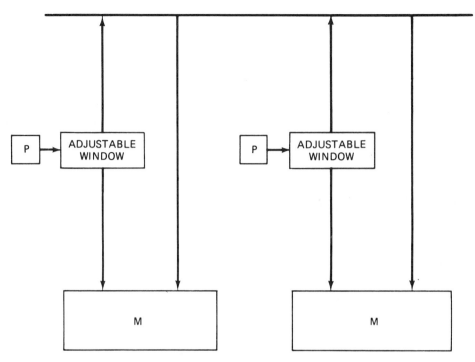

*Figure 9.8: Publicly Accessible Semi-Private Memory*

The major, certainly non-negligible, advantage of this scheme is simplicity.

The second way of assigning private and public memory in a semi-uniform architecture provides for public and private memory with flexible boundaries. This architecture is shown in Figure 9.7. The difference from Figure 9.6 is in the adjustable window logic. It results in a more universal processor board which may be used for applications ranging from purely homogeneous systems with shared memory, to systems having only private memory. An advantage is that boundaries need not have equal settings for all processors in the system. The cost is the extra complexity needed in processor modules, and the need for off board extendable private memory. However, transparency is still a problem.

The third way of arranging private and common memory within a semi-uniform architecture is by far the most flexible solution, but also the most complicated. It provides publicly accessible, semi-private memory. This architecture is shown in Figure 9.8. It requires dual port memory. Each memory board is considered as "belonging" primarily to one processor, but is made accessible to all others. One of the memory ports is connected directly to the "owner" processor's private bus. The second port is connected to a "system" bus to which one or more other processor modules are connected. Each processor module has a flexible window into this public bus. Simultaneous public and private memory accesses are arbitrated by the dual port memory.

Logically, in Figure 9.8, processors each have their own private memory, but may access desired sections of any other processor's private memory by setting a window and using the system bus. This architecture supports a large variety of system configurations, including memory common to all processors, memory common to some processors and strictly private memory within the same system. Aside from the improved flexibility, a major advantage of this scheme is that system buses will unlikely become bottlenecks.

With proper design of the virtual machine, the architecture of Figure 9.8 can be made transparent to the number of processors.

The second major issue concerns the method by which I/O devices are handled within a uniform architecture. I/O devices can be located on a common system bus, on private processor buses extending from the processor boards or directly on the processor boards themselves. All configurations have advantages and disadvantages. Without going into details, we may say that these advantages and disadvantages in many ways are similar to those associated with the various memory configurations.

Based on the foregoing discussion, we may generate a list of key features for a flexible, transparent hardware architecture:

1. A flexible private/public boundary for all memory and I/O. (This feature is needed to ensure generality and easy system growth).

2. External access to the private buses of processors. This feature is needed to extend the private environment of a processor with the desired degree of freedom.

3. A "system bus" structure for all private, public and peripheral card buses. This feature is needed to ensure compatibility among boards, and allows usage of the same boards for private or public purposes. The aim here is to be able to assemble a large number of different architectures using a small number of standard, compatible, general purpose boards.

Given these three essential features, possible architectures range from single processors to homogeneous multiple processors with common memory and I/O only, to various combinations of common, semi-private and strictly private memory and peripherals, all of which can be set up to be fully processor-transparent to the software.

### 9.2.3   Thin-Wire Coupled Architectures and  Reliability

Thin-wire coupling enables individual processing nodes to protect themselves from failures in communications links and in other processing nodes. The question is, can and should thin wire software techniques be used universally? The answer to "can they" is probably yes. The answer to "should they" is probably no, based on experience in the authors' laboratory. This section proposes a mixed approach, which abandons the goal of full transparency and which has proved successful in practice.

A method for including reliability techniques in a large distributed system is shown in Figure 9.9. It is assumed that the system is basically loosely coupled. The system is first partitioned into RELIABILITY ZONES. The idea here is that each reliability zone corresponds to a processing node and can be managed by a single virtual machine. Within each reliability zone, a further partitioning in the software recovery zones may be appropriate. Thus major hardware and software failures are not allowed to propagate across the zone boundaries. Therefore processes in different zones must be thin-wire coupled. Because a reliability zone is managed by a single virtual machine, interprocess communication within its boundaries is fast. Interprocess communication across reliability zone boundaries is expected to be slower and will often involve serial links, with appropriate protocols.

### 9.2.4  Conclusions

This section has attempted to expose by example and discussion some of the issues affecting multiprocessor transparency. The main conclusions are as follows:

— Processor Transparency is achievable with memory-coupled software and hardware architectures.

— Processor and interface transparency are both achievable in principle with thin-wire coupled software architectures, but the issues are complex and efficiency is a potential problem.

— Recommended for distributed systems is a mix of memory-coupling within pre-defined reliability zones and thin-wire coupling between such zones.

— Techniques for structuring memory-coupled and thin-wire coupled systems are discussed. Memory coupling may use monitors (or monitor-like structures) in software and semi-uniform architectures in hardware. Thin-wire coupling requires communications protocols.

— Kernels implement the virtual machine which provides processor-transparency. In general separate virtual machines define separate reliabililty zones which are not transparent to the software.

### 9.3  Efficiency

Kernels reside at the sensitive interface between the hardware and the software where efficiency may be strongly affected by relatively small design refinements and changes. It is the purpose of this section to explore some possible refinements.

First, in Section 9.3.1, we consider kernels which must operate using queues in a time-slicing environment. We call these "queueing kernels". Monolithic, partitioned and distributed kernels of this type are described and discussed. Then some approaches

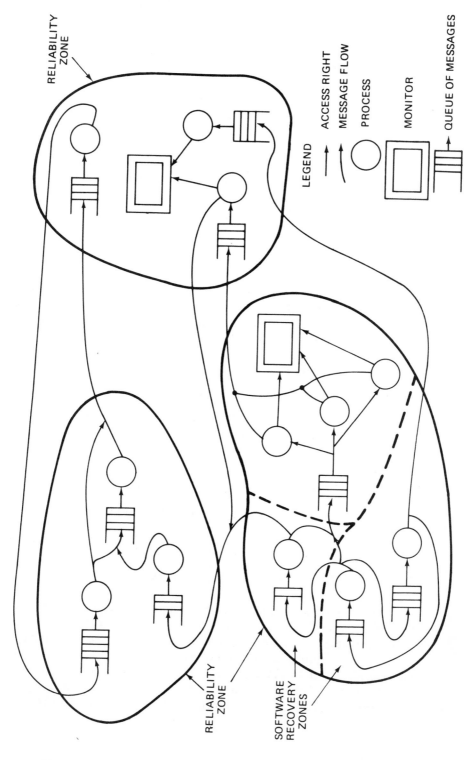

Figure 9.9: Example of Reliability Zones and Software Recovery Zones

**260**

that can be taken at the kernel level to minimize CPU idle time in a time-slicing environment are explored. Concluding remarks on queueing kernels are then presented.

Then in Section 9.3.2 we consider kernels which operate in an environment where process switching normally occurs only by "natural preemption" and not by time-slicing. This environment enables us to introduce the idea of a "polling kernel" in Section 9.3.2. Appendix A supplements this idea by providing a detailed design of an efficient polling kernel based on many of the ideas presented in Sections 9.3.1 and 9.3.2 as well as on other ideas presented throughout the text. This kernel is not presented as the final answer, but only as a concrete example illustrating the ideas.

## 9.3.1  Queueing Kernels

In a time-slicing environment a process switch may be requested at any time by a clock ISR. This leads naturally to a design decision to have the kernel manipulate queues of waiting processes for all calls from any source, as the source is unknown and its time of calling unpredictable. Given that the kernel must manipulate queues, various steps to make it efficient are possible., First we consider relatively inefficient, "monolithic" kernels of the type described in Chapter 5.

### 9.3.1.1  *Monolithic Queueing Kernels*

a) *Characteristics of Monolithic Queueing Kernel*

Consider a kernel which implements the following types of "primitives" (as described in the kernel example of Chapter 5):

— PREEMPT the currently running process

— WAIT or SIGNAL on a sempahore

— BLOCK the calling process or UNBLOCK a specific process in a specific monitor

— Change PRIORITY or "hop" to another PROCESSOR

— Various INITIALIZATION calls as necessary

The collection of data described in Chapter 5 required to support these primitives is here called the Kernel STATE (as distinct from the Process STATE).

The process state transitions followed by processes are shown in Figure 9.10, together with an indication of the kernel STATE components affected by each transition. Not all Kernel STATE components are affected by all transitions. And each state component may be partitioned into components affected only by certain processors. However, in a monolithic kernel, all components are stored centrally, as shown in Figure 9.11.

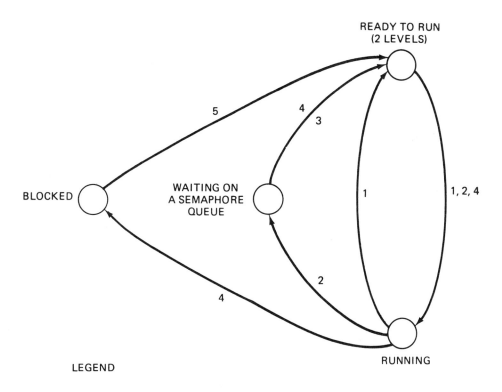

LEGEND

STK — PROCESS STACKS
PD (OR PCB) — PROCESS DESCRIPTION
        BLOCK (OR PROCESS CONTROL BLOCK)
RTR — READY TO RUN QUEUES
SEM — SEMAPHORES


KERNEL PRIMITIVES WHICH AFFECT THE KERNEL STATE
                    (STATE VARIABLES AFFECTED)


1 PREEMPT (STK*, PD*, RTR)

2 WAIT (STK*, PD*, RTR, SEM)

3 SIGNAL (RTR, SEM)

4 BLOCK AND SIGNAL (STK*, PD*, RTR, SEM)

5 UNBLOCK (PD, RTR)


* OF CALLING PROCESS, WHEN SUSPENDED

*Figure 9.10: Kernel State Diagram*

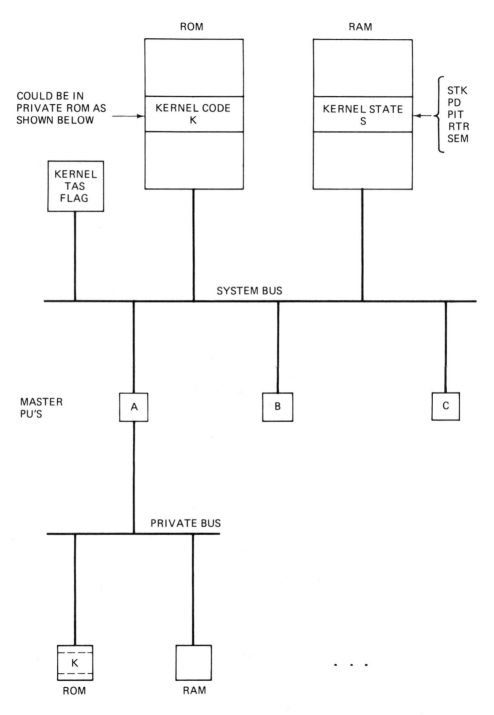

*Figure 9.11: Monolithic Software Kernel*

As described in Chapter 5, access to a monolithic kernel proceeds basically as follows:

Disable interrupts
Lock kernel TAS flag (Test-and-Set flag)
Save registers in the calling process stack
Switch to the kernel stack
Select the kernel primitive
Perform the primitive operation
Switch to the new process stack
Restore the registers from this stack

Interrupts must be disabled first to prevent deadlock due to an interrupt service routine attempting to access an already locked kernel. Locking the TAS flas includes looping on a locked flag.

A monolithic software kernel has certain advantages, as follows:

— state information storage space can be minimized if it is stored centrally;

— overall code storage space can be minimized by keeping kernel code in common memory;

— debugging is simplified by the centralization of system state information: global event traces can be easily maintained; kernel and trace information can be examined by an extra debugging processor connected to the common bus.

It also has certain disadvantages, as follows:

— Execution time is slow because it runs on the calling microprocessor.

— Bus interference while executing the kernel increases the effective kernel execution time.

— Due to the above pair of factors, excessive TAS flag looping may be required by processors requesting the kernel, thus causing further bus interference and delay.

— Due to the requirement for strict mutual exclusion on execution of kernel primitives by different processors, it is not possible to overlap non-interacting kernel calls or parts of calls.

— Due to the fact that both the locking of the TAS flag and the execution of the kernel are performed with interrupts disabled on the calling processor (to avoid kernel calls from interrupt service routines) the kernel places a lower limit on the response time to real time events.

— Execution times may be larger than necessary because of the need for shared code and centralized data.

b) *Improvements to a Monolithic Queueing Kernel*

Within the framework of a software kernel, there are several improvements which can be made:

1. Allow high priority interrupts of the kernel for critical real time events, with no kernel calls allowed from the corresponding interrupt service routines; if consequential kernel calls are required, these could be arranged by triggering lower priority interrupts from the high priority ISR (interrupt service routine). This technique was illustrated in Chapter 7.

2. If the TAS flag is locked, then enable and disable interrupts as part of the loop on the locked flag, to enable critical real time events to be recognized.

3. Put copies of the busy wait loop on the TAS flag in private memory.

4. Place copies of the kernel code in private memory, thus eliminating kernel instruction fetch bus cycles from the bus traffic; however this has the twin disadvantages of using limited private memory and increasing the total system memory requirement.

5. Provide a fast access method for kernel calls which will not cause process switching, such as SIGNALs on semaphores, or WAITs on semaphores when the queue is empty.

Beyond these items, not much can be done without speeding up the execution of the kernel itself. This may be accomplished by hardware, as described in Section (c) below.

c) *Monolithic Kernel Device*

A monolithic kernel device may be used to speed up the kernel in a tightly coupled system at the expense of greater hardware complexity. This device could employ a fast microprocessor (perhaps a bit-sliced microprocessor).

Some questions to be considered are:

— Is the fast kernel to be capable of being a bus master, so that parameters for the kernel can be placed in common memory, or is it just a slave peripheral? A bus master requires more complex hardware logic.

— Is kernel arbitration to be performed by looping on a TAS flag, or by a special hardware arbiter which mediates wait, hold or halt states? A special arbiter is more efficient but also more costly and complex.

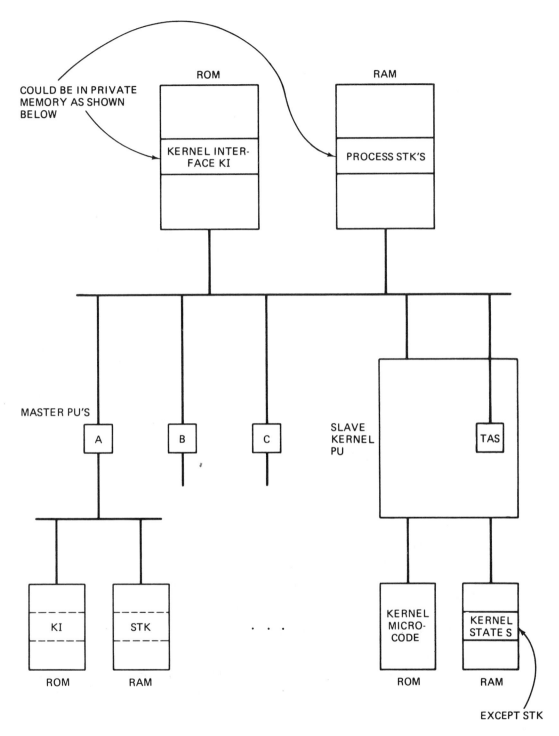

*Figure 9.12: Monolithic Kernel Device*

265

— Should advantage be taken of existing hardware logic to provide the kernel with parameters, such as the calling processor identity, which would otherwise have to be passed explicitly by software? Hardware modularity is reduced by this approach, for the sake of a small gain in performance.

— What LSI components should be used to implement the fast kernel?

A system with a fast kernel device is shown in Figure 9.12; it is a slave on the system bus, is protected by a TAS flag, receives all parameters under software control over the system bus and could use a bit-sliced microprocessor. Indications are that such a kernel could, with a bit-sliced microprocessor, provide an order of magnitude speed improvement over the same kernel executed by an INTEL 8080. Faster processors than the 8080 would proportionally reduce this gain in speed.

The speed improvement would be greater were it not for the interface code required. The calling processor must still maintain a set of process stacks. Therefore, in addition to the overhead of passing the usual kernel parameters via I/O instructions, the calling processor must still perform all stacking and unstacking operations to save and restore the process state. The kernel only saves stack pointers passed to it by the calling processors. Furthermore, the calling processor must still loop on the kernel TAS flag.

Because kernel execution time is a known quantity (there is no interleaved processing form other processors, as there is with a software kernel), it is sufficient for the calling process to proceed as follows:

```
Disable interrupts
Lock TAS
Save registers
Write parameters directly to kernel (not to common memory)
Execute short timing loop
Read result from kernel
Restore the stack pointer
Restore registers from the stack
Unlock TAS
Enable interrupts
```

There is no need either to go to a HALT state followed by an interrupt when the kernel is done, or to poll the kernel for results.

d) *Multiport Monolithic Kernel Device*

The TAS flag can even be completely eliminated by using a multiport kernel in which each processor has its own private entry port and parameter buffer memory as shown in Figure 9.13. The kernel is then responsible for servicing requests one at a time in FIFO order.

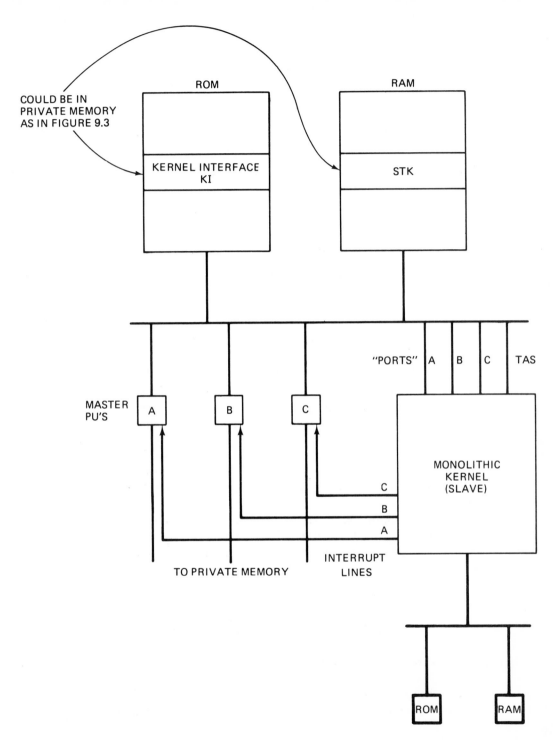

*Figure 9.13: Multiport Monolithic Kernel Device*

This approach has the disadvantage that kernel execution time as seen by the processor is now a variable which depends on the number of processors in the kernel queue. Therefore, either

— polling for kernel-done is necessary, (which has the same disadvantages as polling for kernel available by looping on the TAS flag), or

— as shown by Figure 9.13 processors halt after writing parameters to the kernel and are interrupted by the kernel when done by a special high priority interrupt (which has the disadvantage that interrupt lines are required from the kernel processor to every CPU).

### 9.3.1.2   Kernel Classifications

To avoid the disadvantages of monolithic kernels, the kernel STATE may be partitioned and distributed (at design time) as follows:

a) Partitioned: Fragment the kernel state so that concurrent activity may be allowed on different fragments. Partitioning may be performed horizontally (by processor) or vertically (by memory level). Partitioning can provide increased concurrency in a physically tightly coupled system.

b) Distributed: Place the fragments of a partitioned kernel state entirely under the control of individual processors. With a distributed kernel only a small amount relatively infrequently accessed data must be shared between processors.

A discussion of partitioned and distributed kernels is given in Sections 9.3.1.3 and 9.3.1.4. First we examine some basic issues concerning kernel partitioning and distribution.

Clues on how to perform the partitioning and distribution are provided by Figure 9.10 which shows which kernel state fragments are affected by which primitives. There are three basic questions which must be answered in any partitioning or distribution of the kernel:

1.  How is mutual exclusion to be implemented?

2.  Where is the kernel state information to be stored?

3.  How is the kernel logic to be implemented?

As shown in Table 9.1, kernel types may be classified as Monolithic, Partitioned, or Distributed, depending on how these questions are answered. Each question is considered in turn below:

### a) Mutual Exclusion

IN all cases the basic mutual exclusion mechanism on the fragments of the kernel state is assumed to be provided by one or more indivisible, shared test-and-set (TAS) flags. In a tightly coupled environment, a TAS flag may be provided by a TAS instruction, by a suitable read-modify-write instruction which can mimic a TAS operation, by a sequence of instructions executed while the system bus is locked, or by a separate shared hardware peripheral.

### b) Location of Kernel State

Mutual exclusion may also be accomplished, in a distributed (type 3) kernel, by physical isolation of state information in private memory. If activities on other processors are to be allowed to affect this state information (as they must in certain cases), then some limited intra-kernel, inter-processor communication mechanism must be provided through a small number of shared flags and/or counters; the method of communication may limit the synchronizing capability of the kernel, (Type 3A).

The only choices for location of kernel state fragments are in shared memory (Types 1A, 2A), in private memory (Types 1B, 3A, 3B) or partly in both (Type 2B).

### c) Implementation of Kernel Logic

Kernel logic may be implemented as software running only on the calling processor (Types 1A, 2A, 2B), as software running in a distributed fashion on more than one processor (Types 3A, 3B) or as a separate hardware device (Type 1B).

#### 9.3.1.3  *Partitioned Kernels (Type 2)*

Partitioned kernels can improve tightly coupled systems by reducing common memory requirements and increasing concurrency. In partitioned kernels, the rule of mutual exclusion on the execution of entire kernel primitives is relaxed. Components of the system state such as semaphores and RTR queues are selectively protected by test and set (TAS) flags. Deadlock is avoided by ensuring that processes release a TAS-protected resource before requesting another and that interrupt service routines which call the kernel are not allowed to interrupt kernel execution.

#### (a)  *Horizontally Partitioned Kernel (Type 2A)*

The general features of a horizontally partitioned kernel are summarized in Table 9.1. Detailed features are as follows (see Figure 9.14):

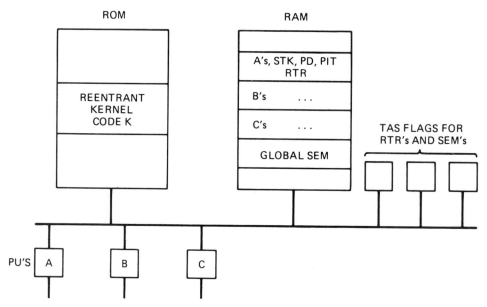

*Figure 9.14: Horizontally Partitioned Kernel (Type 2A)*

— each RTR queue and semaphore is protected by a TAS flag (or flags could be grouped, for economy, with some loss of concurrency);

— queues are linked lists of process descriptions, in common memory, accessible to all processors;

— kernel code outside the TAS-flag-protected critical sections is reentrant so that it can be concurrently shared by different processors; and

— there is a separate kernel stack for each processor.

With this type of arrangement, the kernel can be executed completely concurrently by different processors, except for the (rare) occasion when more than one processor is attempting at the same time to access the same TAS-flag-protected semaphore or RTR queue.

However, all system state information is still in common memory accessible by all processors.

A kernel call now proceeds as follows (illustrated by a SIGNAL):

```
Disable interrupts
Lock semaphore TAS
Do semaphore operation to get (perhaps) a pointer to the process
description of a process to be awakened
Unlock semaphore TAS
Lock RTR queue TAS
```

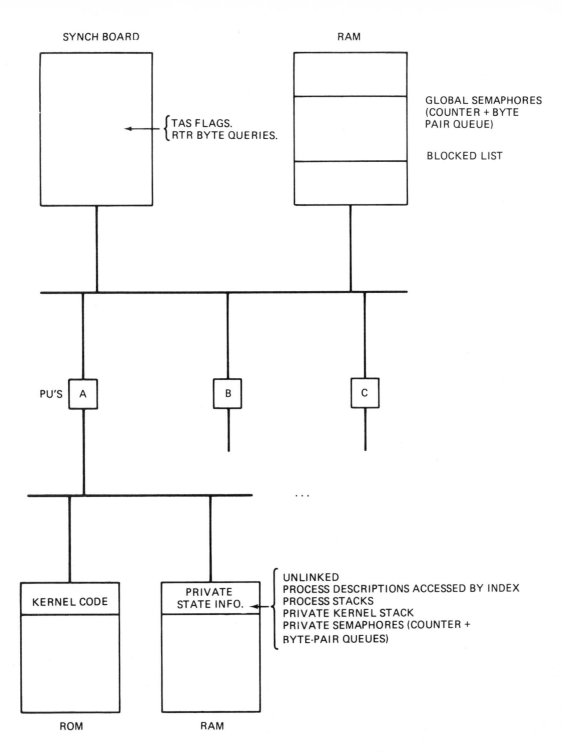

SYNCH BOARD

RAM

GLOBAL SEMAPHORES
(COUNTER + BYTE
PAIR QUEUE)

BLOCKED LIST

{ TAS FLAGS.
  RTR BYTE QUERIES.

PU'S   A          B          C

. . .

KERNEL CODE

PRIVATE
STATE INFO.

{ UNLINKED
  PROCESS DESCRIPTIONS ACCESSED BY INDEX
  PROCESS STACKS
  PRIVATE KERNEL STACK
  PRIVATE SEMAPHORES (COUNTER +
  BYTE-PAIR QUEUES)

ROM          RAM

*Figure 9.15: Vertically Partitioned Kernel (Type 2B)*

271

**Table 9.1 — Kernel Classifications**

| Kernel Type | Subtype | Characteristics | | |
| --- | --- | --- | --- | --- |
| | | Mutual Exclusion | System State | Kernel Logic |
| 1. Monolithic | A Software | Kernel primitives are strictly mutually exclusive for all processors, enforced by a test-and-set (TAS) flag | In common memory accessible to a processors | Executed by the calling processor. Code may be duplicated in private memory or shared in common memory |
| | B Separate Hardware Device | as for Type 1A | In private memory of a special fast slave processor | Executed by a special fast slave processor from its private memory |
| 2. Partitioned | A Horizontal (by processor) | Data is selectively locked. Primitives may be executed concurrently by several processors as long as each operates on separate data; enforced by TAS flags | Horizontally partitioned in common memory into data associated with particular processors and global data | Executed by the calling processor. Code may be duplicated in private memory or shared in common memory |
| | B Vertical (by memory level) | As for Type 2A | As for Type 2A except data is also vertically partitioned to minimize common memory requirements. Queues are still in common memory, though in reduced form | As for Type 2A |

| | | | |
|---|---|---|---|
| 3. Distributed | A<br>Restricted Synchronization Capability | Processors may not access private kernel data associated with other processors; in particular they may not access queues | as for Type 2A except most data is in private memory, including queues. Only a few flags and counters are in common memory. Semaphore queues belong to particular processor; semaphore waits may only be performed by the processor owning the queue | Executed in the first instance by the calling processor but with need for consequential action in some cases by other processors. Kernel Code is Duplicated in the private memory of each processor |
| | B<br>Unrestricted Synchronization Capability | As for Type 3A | 3As for Type 3A except semaphore queues are horizontally distributed; semaphore waits may be performed by any processor | As for Type 3A |

```
Add process description to the RTR queue
Unlock RTR queue TAS
Enable
Return
```

### (b)  *Vertically Partitioned Kernel (Type 2B)*

To reduce common RAM memory requirements, a further partitioning of the kernel is possible, in a "vertical" direction.

Here all queues in common memory are reduced to queues of indices or counters, which are used by a processor to identify larger data elements stored in the private memory of individual processors.

Thus a RTR queue for a particular processor is a FIFO byte queue of process indices. A semaphore is counter together with a FIFO queue of process i.d.'s, where process i.d. is a combination of process index and processor index. The actual process descriptions are in the private memory of the processor "owning" the processes and are accessed by process index. Note that with this arrangement processes are firmly dedicated to processors; processor "hopping" is not allowed.

The possibility exists, with this arrangement, of using FIFO chips for the RTR byte queues instead of RAM. Then RTR queues do not need TAS flag protection because byte queueing operations on FIFO chips can be accomplished in one bus cycle. Semaphores, however, need to be in RAM, with TAS flag protection.

The following arrangement then becomes possible (Fig. 9.15):

— the RTR queues and the TAS flags are grouped on a single "synch" board on the system bus;

— semaphores may be *global* or *local*; global semaphores, at least, are in common RAM;

— special semaphore operations *cross-wait* and *cross-signal* are defined on global semaphores which differ from monolithic semaphore operations only in their use of the TAS flags for locking the semaphore;

— all semaphore operations, whether global or local, require interrupt lockout protection, although this can be relaxed somewhat if sources which are allowed to interrupt the kernel do not call the kernel.

With this arrangement, special kernel routines for global semaphores are as follows:

*CROSS SIGNAL* (on a global semaphore)

```
Disable interrupts
Lock semaphore TAS
Do global semaphore operation to get (perhaps) process i.d.
Unlock semaphore TAS
```

```
Write process index into appropriate RTR queue, (leave it to
the time-slice interrupt routine on the processor owning that
queue to dispatch the process via a PREEMPT call if no one is
running)
Enable interrupts
Return
```

*CROSS WAIT* (on a global semaphore)

```
Disable interrupts
Lock semaphore TAS
Do public semaphore operation, queue process i.d. if
suspension
Unlock semaphore TAS
Save registers; read new process index from RTR; restore new
process registers
Enable interrupts
Return
```

### 9.3.1.4 Distributed Kernels (Type 3)

In a distributed kernel the kernel state is almost completely private to a processor. Not even reduced versions of the queues are kept in common memory as in Type 2B. Thus a processor's RTR and semaphore queues cannot be accessed by another processor and a different mechanism is required for cross waits and signals. This approach holds promise in at least the following two ways:

1. Reducing bus traffic in a system with heavily used common memory or common devices on the system bus.

2. Enabling the use of existing programs, which use the kernel to implement Concurrent High Level Language constructs, in a new environment where there is very limited and possibly slow shared memory; in such an environment only a limited number of status bytes, counters and parameters can be stored in shared memory.

### (a) Cross-Processor Communication

The problems inherent in distributed kernel can be illustrated by the example of semaphore WAIT and SIGNAL operations. Two cases can be distinguished:

### 1. Restricted Synchronization Capability

Waits on a particular semaphore are only performed by processes running on one processor; cross signals can however be performed from other processors. This case arises when processes simply wish to wait on events signalled by processes running on other processors.

2. *Unrestricted Synchronization Capability*

Several processors may wait on the same semaphore. This case arises, for example, if monitors are to be shared between several processors so that a monitor gate semaphore is required which is accessible from all processors.

(b) *Distributed Kernel with Restricted Synchronization Capabilities (Type 3A)*

In the first case each processor could easily manage its own ready-to-run queues and the queues of all semaphores on which it may wait. It would always be responsible for transferring processes from the semaphore queue to its own ready-to-run queue after a signal by any process. Signals on the semaphore from the same processor are handled in the usual way by standard kernel call. However, signals from other processors must be handled differently because such processors are not allowed to access the appropriate queues.

One approach to this problem is to implement semaphores as follows:

— semaphores are "owned" by processors

— a semaphore queue is linked lists of process descriptions in the private RAM of the processor

— the semaphore *state* consists of:

— a *wait counter* which may take any value, including negative values; recall that with ordinary semaphores if this counter is negative, its absolute value indicates the number in the queue; now it indicates the number of processes eligible for wakeup

— a *wakeup counter* which takes only non negative values; it indicates the number of wakeups which have been performed without yet removing processes from the private semaphore queue. The wakeup-counter is required because negative values of the wait-counter indicate how many processes are left in the queue only if the wakeups are done immediately. With CROSS-SIGNALS wakeups are deferred and, in fact, several may be accumulated for later action; the number accumulated is recorded in the wakeup-counter.

Then when one or more processes are queued on the semaphore, as indicated by a negative value of the wait-counter, a cross-signal primitive simply increments the wakeup-counter for later examination by the time-slicing interrupt service routine on the processor owning the semaphore. There is then an inherent latency of one time-slice in the cross-signal operation.

The system structure is shown in Figure 9.16; the algorithms are as follows:

Special WAKEUP performed by Clock ISR on processor owning the semaphore, for every semaphore so owned

```
Disable interrupts
Lock semaphore TAS
```

```
If wakeup counter > 0 , then
    Begin
    Temp ← Wakeup-counter
    Wakeup-counter ← 0
    Unlock semaphore TAS
    Transfer Temp processes to RTR queue
    End
Else unlock semaphore TAS
Enable interrupts
```

Normal WAIT executed by processes on the processor owning the semaphore queue.

```
Disable interrupts
Lock semaphore TAS
Decrement wait-counter
Temp ← wait-counter
Unlock semaphore TAS
If Temp < 0 then suspend current process on the semaphore queue
and schedule another process
Enable interrupts
```

Normal SIGNAL executed by processes on the processor owning the semphore queue.

```
Disable interrupts
Lock semaphore TAS
    If wait-counter < 0 then transfer one process from the
    semaphore queue to the RTR queue
Increment wait-counter
Unlock semaphore TAS
Enable interrupts
```

Special CROSS-SIGNAL performed by processes not having access to the semaphore queue

```
Disable interrupts
Lock semaphore TAS
If wait-counter < 0 then increment wakeup counter
Increment wait-counter
Unlock semaphore TAS
Enable interrupts
```

(c) *Distributed Kernel with Unrestricted Synchronization Capability (Type 3B)*

In the second case as illustrated by Figure 9.17, every processor maintains a semaphore queue for all global semaphores. CROSS-WAITS are now identical to ordinary WAITS; the process simply waits in the queue on its own processor. CROSS-SIGNALS are as before, WAKEUPS are performed in a similar fashion, but now every global semaphore must be polled for wakeup-count > 0 (instead of only those

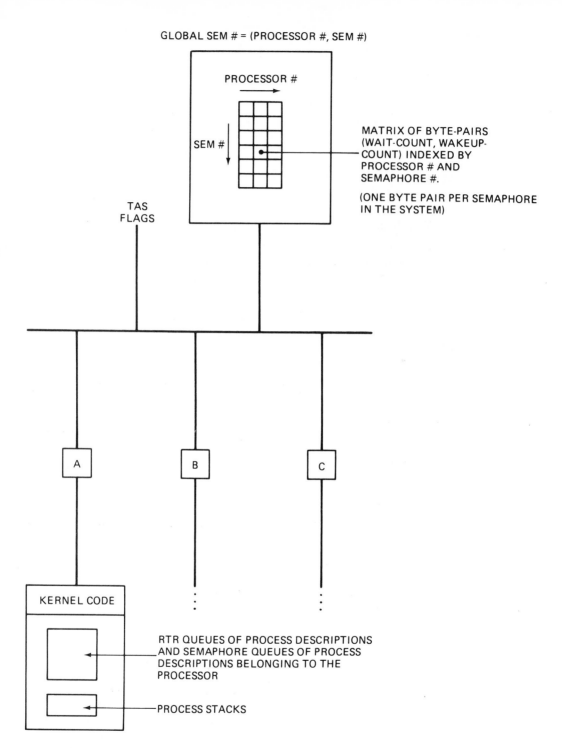

GLOBAL SEM # = (PROCESSOR #, SEM #)

PROCESSOR #

SEM #

MATRIX OF BYTE-PAIRS (WAIT-COUNT, WAKEUP-COUNT) INDEXED BY PROCESSOR # AND SEMAPHORE #.

(ONE BYTE PAIR PER SEMAPHORE IN THE SYSTEM)

TAS FLAGS

A

B

C

KERNEL CODE

RTR QUEUES OF PROCESS DESCRIPTIONS AND SEMAPHORE QUEUES OF PROCESS DESCRIPTIONS BELONGING TO THE PROCESSOR

PROCESS STACKS

*Figure 9.16: Distributed Kernel with Restricted Synchronization Capability (Type 3A)*

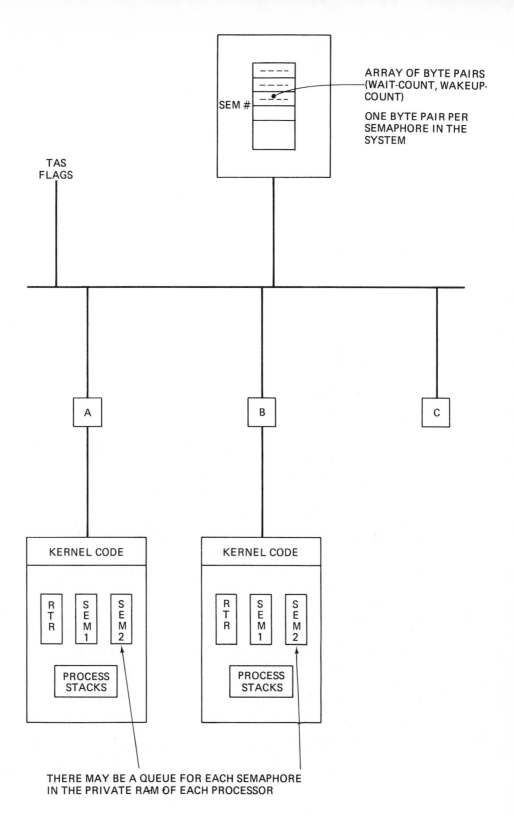

ARRAY OF BYTE PAIRS
(WAIT-COUNT, WAKEUP-COUNT)

ONE BYTE PAIR PER
SEMAPHORE IN THE
SYSTEM

SEM #

TAS
FLAGS

A

B

C

KERNEL CODE

R T R    S E M 1    S E M 2

PROCESS
STACKS

KERNEL CODE

R T R    S E M 1    S E M 2

PROCESS
STACKS

THERE MAY BE A QUEUE FOR EACH SEMAPHORE
IN THE PRIVATE RAM OF EACH PROCESSOR

*Figure 9.17: Distributed Kernel with Unrestricted Synchronization Capability*
*(Type 3A)*

"owned" by the processor) and the number of wakeups performed by any processor for any semaphore is bounded by the number of processes in its own queue for that semaphore.

A consequence of this organization is that the semaphore queue discipline is random service because wakeups will be performed asynchronously by different processors.

### 9.3.1.5   *Improved Methods of Handling CPU Idle Time*

At various times the ready to run queue(s) of a processor are empty; all processes are suspended on semaphores, condition variables, etc., and an external signal from a hardware interrupt or another processor is required to get things moving again. Such intervals are called CPU idle time. This section addresses the question of what to do with a processor during its idle intervals.

A common technique for handling idle time on a uniprocessor is to dispatch a special idle process. Typically the idle process loops on a PREEMPT kernel primitive in order to allow other processes to run as soon as possible after being signalled. This method, however, is impractical on a multiprocessor since the presence of one or more idle CPUs would cause intolerable kernel and system bus congestion. One solution adopted is to let the idle process do something innocuous, such as execute a tight loop out of private memory or go into a halt state, until the clock ISR finally issues a PREEMPT to time slice it.

The problem with the above idle time technique is the latency of up to one clock period or one half clock period on average before a process placed on the ready-to-run queue during an idle period has a chance to run. This occurs, for example, in semaphore SIGNAL primitives. The latency may cause problems other than low efficiency; for example, excessive transmission delays in a packet switching package may result. Consider the DLC/ISR/CRC package described in Chapter 7 and suppose that the DLC module runs on one processor and the ISR/CRC module on another. Delays in "hopping" between the modules on one processor and those on another may result in occasional packet loss, which requires a much longer time for recovery than just the latency period.

Clearly, reducing the clock period is a limited remedy at best, because at some point PREEMPT calls would be too frequent. Several processors, each generating PREEMPTs at the higher rate, would greatly increase the queueing delay seen by another processor trying to make more legitimate use of the kernel.

Two alternative solutions due to Cavers and Bezanson are introduced here. Neither requires much change to the kernel level software and both reduce processor latency to a period of the order of the time required for a kernel call.

### *Method 1: Wires*

A parallel interface with one bit per processor is required to be publicly available on the system bus. A wire is taken from each bit on the public parallel interface to a special bit on the private parallel interface of the appropriate processor. The scheduler part of the kernel is modified so that:

— a processor sets its bit in the public interface whenever it has to run the idle process;

— whenever a process is placed on any ready to run queue of a processor the corresponding bit in the public parallel interface is cleared.

The idle processes on all processors are identical. They simply loop on the special bit of the private interface until it is cleared, whereupon they call PREEMPT.

Evidently the idle process suspends itself as soon as another process is ready to run, and only one PREEMPT is issued per idle period. The latency has been cut to the duration of the PREEMPT kernel call. A further advantage is that neither a clock ISR or a hardware clock required, provided all processes run to suspension points without executing polling loops. The wires, however, are a potential problem.

*Method 2: No Wires*

In this method a location in common memory is used instead of the public parallel interface. Again there is a bit associated with each processor, with the same interpretation as in method 1; the bit is set when that processor schedules the idle process, and is cleared whenever a process is placed on any ready to run queue of that processor. The idle process polls this bit at intervals determined by a tight timing loop executed out of private memory, and when the bit is cleared the idle process calls PREEMPT.

The duration of a kernel call would seem to be a suitable interval for the idle process in its polling of the bit. Certainly there is no perceptible increase in system bus load at this rate. We now see processor latency slightly above that of method 1, including the execution time of the PREEMPT. The obvious, advantage is that no public parallel interface and no wires are needed.

### 9.3.1.6  *Concluding Remarks on Queueing Kernels*

(a) *Execution on Different Processors*

Some designs may require a process to "hop" to another processor, that is continue executing on another processor, by making an explicit call on the kernel to change processors. Processor hopping may be used, for example, to access the private single-port memory of another processor. The prime requirement to make it possible is that the process stack must be accessible to both processors (either in shared memory or duplicated in private memory). Partitioned and distributed kernels therefore make processor "hopping" difficult or impossible.

(b) *Performance*

As one motivation for considering kernel types other than Type 1A (Monolithic software), consider the example discussed in Chapter 5 of an X.25 protocol on a I808A system with a monolithic software kernel, where approximately 20 kernel calls are required to send or receive a single packet. In that system there is a kernel overhead of at least 15% for a 3-processor system running one X.25 channel at 9600 baud, with an additional overhead of at least 10% for every additional channel running at the same rate.

## 9.3.2   Polling Kernels

As a final step towards kernel efficiency, we introduce the idea of a polling kernel. Such a kernel does not provide for time slicing (except as an emergency measure) and thereby enables advantage to be taken of the fact that only the processes themselves can trigger a process switch.

First some remarks are made in Section 9.3.2.1 about clock preemption. Then in Section 9.3.2.2 polling is considered as an alternative to queueing and appropriate algorithms for a polling kernel are described.

### 9.3.2.1   About Clock Preemption

It may be argued that clock preemption is "fair" in the sense that all system processes get their "fair" share of processing time. This point is certainly true in a user-programmable time sharing environment. However, in a special purpose, real-time environment, the same "fairness" argument may become less relevant. Let us look at the cost of clock preemption:

— Preemption is an inefficient algorithm in any kernel because of the amount of computation associated with process switching. This point becomes increasingly true as the number of system state variables which must be saved and restored on each process switch increases.

— Clock preemption creates an environment where anything can happen at any time. Without going into details it may be said that extra precautions and protection mechanisms are necessary for such an environment.

— Clock preemption has a tendency to cause indirectly other process switches. Whenever processes are preempted within monitors, the gates are closed for larger than normal periods of time with the result that more processes than normal must wait on monitor gates and more process switches than normal must occur.

— On the average many clock preempts must occur in the vicinity of where natural preempts would have occurred anyway causing unnecessary process switching.

— The overhead of attempting to preempt a process still exists regardless of whether another process is ready to run or not.

Based on these points and on our experience with the nature of special purpose applications, it is suggested that except for a safety valve discussed below, it may often be sufficient to allow only natural preemption using one of the following mechanisms:

1)  WAITS (on semaphores, gates, condition variables, etc.);

2)  DELAY, SUSPEND, SLEEP or STALL primitivves;

3)  A PREEMPT primitive to be used by the programmer whenever it seems desirable to guarantee that no more than a given number of high level language statements will be executed.

To ensure that no single process consumes all the processing time even under extreme load conditions, a safety valve, namely a TIMEOUT PREEMPT mechanism, is suggested, triggered by a clock ISR. For some systems, TIMEOUT PREEMPTION could exist at debug time only to determine where it occurs and to install calls to natural preempts whenever appropriate. Such a mechanism may also be used to obtain statistics on maximal execution times for system processes, by determining for what minimal value of timeout counts clock preempts do not occur.

### 9.3.2.2 *Polling as an Alternative to Queueing*

Partitioned or distributed kernels are particularly efficient if one simplifies the data accesses and the data access sequences associated with individual kernel primitives. Doing this implies making modification to monolithic kernel data structure concepts. The goals to aim for are to have frequently executed primitives perform simple short data structure manipulations, to avoid indivisible sequences of operations on several data structures and to allow concurrent manipulations of unrelated data structures. This section proposes one way of achieving this. The proposal is based on removing the various queueing manipulations associated with all primitives and replacing them by polling in the process switch algorithm only. There is a substantial advantage to be gained if those primitives do not result in a process switch, at the expense of less efficient process switching.

Let a semaphore consist of two counters as for the distributed kernel of Section 9.3.1, a wait count and a signal- count. However the counters here have different interpretations and are used differently because of the absence of queueing. Signalling corresponds to incrementing the signal-count and nothing else and hence is always efficient. Waiting corresponds to incrementing the wait-count and then executing the following algorithm:

```
IF WAITCOUNT- SIGNALCOUNT = POSITIVE THEN PROCESS SWITCH
```

Waiting is thus always efficient when process switching is not involved. Process switching requires saving the semaphore name and the current wait-count as part of the process description. Then the following holds true for any non running process at the time of a process switch, where BARRIER$COUNT is the saved wait-count and SEM.SIGNAL$COUNT is the signal-count of the saved semaphore:

```
IF FOR PROCESS DESCRIPTION
    BARRIER$COUNT-SEM.SIGNAL$COUNT = POSITIVE
    THEN PROCESS IS NOT READY TO GO
    ELSE PROCESS IS READY
```

Hence, it is possible for a preempt routine to poll process description blocks until one is found which is able to continue. In effect, the queueing associated with a WAIT (CONVENTIONAL$SEM) and the dequeueing from the semaphore queue and queueing onto the ready to run queue associated with a SIGNAL (CONVENTIONAL$SEM) has been replaced by a polling algorithm in the process-switching routine. Preemption efficiency thus depends on the efficiency of the polling algorithm but in any case is likely to be less than for queueing kernels if there are many processes. Detailed algorithms are given in Figure 9.18.

Process switching is undoubtedly the least efficient algorithm of the kernel. Because of that, it is preferable to minimize the amount of preemption in the system. Note however that other processors may execute any primitive entirely concurrently with a preempt on one processor. Moreover, interrupts remain on during preempts (except for clock interrupts), and the inefficiencies associated with preempt are thus not critical as far as time critical ISRs are concerned.

The following remarks apply to Figure 9.18:

— The kernel contains no data structures. Instead, semaphore counts and process description blocks are declared privately and passed as parameters for all kernel calls.

— Except for the critical regions the kernel is entirely concurrently executable. (This remains true as other primitives are added.)

— Clock preempts are not forbidden but are inefficient due to polling.

— Critical regions are never long. (This remains true as other primitives are added.)

— Signal operations are sufficiently short to be performed by time critical ISRs.

— Very little code is associated with primitives. (This remains true as other primitives are added.)

— Other than semaphore counts for semaphores used by processes on different processor boards, no data structures must reside in common memory.

```
SIGNAL:  PROCEDURE (SEM$COUNTS$AT);
         DISABLE;
         LOOP ON TEST AND SET FLAG;
         INCREMENT SEM.SIGNAL$COUNT;
         RESET TEST AND SET FLAG;
         ENABLE;
END SIGNAL;
WAIT:    PROCEDURE (SEM$COUNTS$AT, PROC$DES$BLK$AT);
         DISABLE;
         LOOP ON TEST AND SET FLAG;
         INCREMENT SEM.WAITCOUNT;
         SEM$STATE = SEM.WAITCOUNT - SEM SIGNAL COUNT;
         RESET TEST AND SET FLAG;
         ENABLE;
         IF SEM$STATE=POSITIVE
         THEN CALL PROCESS$SWITCH;
         RETURN;
END WAIT;


PROCESS$SWITCH:  INTERNAL PROCEDURE
         MASK CLOCK INTERRUPT
         SAVE WAIT$COUNT , SEM$INDEX AND SYSTEM STATE ON
             PROC$DES$BLOCK;
LOOP:    LOCATE AND LOAD NEXT PROCESS DESCRIPTION BLOCK
         IF BARRIER$COUNT-SEM.SIGNALCOUNT = POSITIVE
         THEN GO TO LOOP
         RESTORE NEW SYSTEM$STATE
         UNMASK CLOCK INTERRUPT
         RETURN
END PROCESS$SWITCH
```

*Figure 9.18: Algorithms for a Polling Kernel*

Based on these and other concepts, a full kernel was designed for a multi-8080 system with the same primitives as those provided by the example kernel of Chapter 5 and some others (see Section 9.3.3). The following encouraging statistics were recorded:

— TOTAL SIZE: less than 200 bytes of code.

— Execution times for all primitives not involving any preempting (signal, continue, attach, restart, successful wait, etc. ...):

less than 50 usec.

— Primitives involving preempts (waits, delays, preempts, etc. ...)

$$\pm\ 70\,\text{us}\ +\ n(70\,\text{us})$$

where n = number of process description blocks polled before eligible process is found.

The single largest drawback is the time it takes to perform preempts. (This time is proportional to the number of processes running on a particular processor).


## 9.4   Testing and Debugging

Testing and debugging of multiple microprocessor systems designed using the CHLL approach of this text proceeds from the bottom up, by testing the following system elements, in the order shown:

1.  Hardware

2.  Low Level I/O

3.  Kernel

4.  CHLL Level

Only the last element requires comprehensive special treatment. The others are handled by more or less conventional techniques, using Microprocessor Development Systems, In-Circuit Emulators and, possibly Logic Analyzers. The proliferation of these expensive items of equipment is a major contribution to high development costs

### 9.4.1   Hardware Testing/Debugging

Hardware is debugged using conventional techniques. As with any system, the availability of a good hardware diagnostic package is an important asset in all phases of system testing and debugging, so that true software errors can be readily identified. This is particularly important for concurrent systems which are prone to mysterious time-dependent bugs which are often difficult to pinpoint as hardware or software problems on the basis of their observed effects. An independent hardware diagnostic can often resolve such problems by revealing a hardware failure or giving the debugger confidence in the hardware.

Logic analyzers and in-circuit-emulaters are valuable tools for hardware testing, debugging and diagnosis. Logic analyzers have the advantage of not requiring a microprocessor development system. However, such a development system, which provides program editing, compiling, linking, running, etc,. as well as in-circuit emulation services is necessary in any case. Indeed, it may be desirable to have access to both types of tools because they allow testing from two different viewpoints; such a two-pronged approach is bound to be more thorough.

### 9.4.2  Low Level I/O Testing/Debugging

The main problem here is usually understanding the documentation describing the operation of the I/O chips (timers, interrupt chips, parallel I/O chips, USART chips, etc.). The computer world is unfortunately split into hardware and software camps and user interfaces and documentation produced by the hardware camp are often difficult and complex to use as the basis for programming. The microprocessor revolution is changing this situation but its effects will still be felt for some time to come.

As a result, the debugging of low-level I/O tends to be a tricky, hands-on, excessively lengthy activity. However it uses the same techniques as for hardware diagnostics.

### 9.4.3  Kernel Testing/Debugging

The kernel is typically written in assembly language to minimize memory and execution time requirements and is usually no more than a few kilobytes of code and data. It is simply a collection of procedures which access common data and as such is easily debugged in isolation by standard interactive debugging techniques using either a primitive console ROM monitor or a microprocessor development system. The main objective is to exercise all code and to verify its effects on the data structures.

One useful aid to kernel testing is as follows: The kernel is designed with a boolean debug flag which causes a print-out of the progress of a call at each critical stage. For example in debug mode the calling process can be identified by index parameters. On leaving the kernel the index and return parameters can be printed. Additional print-outs can be obtained from any primitive or operation. For example the manipulation of queues often causes problems. Debug print-outs can be inserted to record the queue address and the element added or deleted.

Following this basic debugging phase the kernel's ability correctly to interleave concurrent requests must be verified. During this phase it is useful to replace the printout of the kernel activity with a ring buffer which records the last n activities. Following a crash, this buffer can be read directly (in Hex) or printed. This technique is useful for CHLL testing as well. Examination of past activity is often the only way of recreating the condition which caused the crash. This particular kernel feature could be left in a production model to assist in trouble shooting after installation (the feature would normally be disabled).

Kernel traces can dramatically increase kernel execution time and thereby expose problems which are execution-time dependent. For example, the possibility of losing characters in synchronous transmission due to kernel uninterruptability (Section 7.7) was discovered during system testing in the authors' laboratory when the kernel trace feature was enabled. Kernel traces are important but cannot be left enabled permanently.

### 9.4.4   CHLL Level Testing/Debugging

This section describes two approaches to CHLL testing/debugging used in the authors' laboratory on a multimicroprocessor system. The approaches are general and the specific examples are only used for concreteness. The first approach is used for internal software which is not subject to the same stresses as communications protocol software and which is, therefore, easier to test and debug. The second approach is used for communication protocols which are subject to high stresses because of their complex, asynchronous nature and because of the many types of errors which may occur.

#### 9.4.4.1   Internal CHLL Software

Internal software is less subject to stress than communication software. A limited set of debugging tools and techniques is therefore often sufficient, as described below:

1) Looping on Flags if set:

A public procedure which allows the user to loop indefinitely on a specified (as input parameter) flag if set, is useful. This tool replaces the use of BREAKPOINTS, and has the following advantages:

— Flags can be set and reset at will using an "observer" processor on another processor board (multiprocessing has its advantages):.

— It is not necessary to remember the instruction which was replaced by a breakpoint (a problem with debugging using a ROM console monitor).

— It is not necessary to generate cross-reference listings of program lines versus memory locations, as is required for the use of break-points (a problem with debugging using a ROM console monitor).

— Various real time events may be simulated by setting and resetting flags as desired (for example, bottlenecks or breakdowns may be simulated).

— The looping on flags may still take place in ROM based multiprocessor systems whereas the insertion of breakpoints may not.

2) Tracing code sections

Each section (paragraph of code) is numbered and a call to a procedure is made each time a section has been executed. The procedure stores that fact in a ringbuffer belonging to the process running at the time of the call. As such, it becomes possible to know (any time) the last N sections of code executed by each process, thereby facilitating the tracing of past activities.

3) dynamic testing of error conditions, inconsistencies and assertions.

At various points throughout the system modules, literally declared calls are inserted which check for error conditions, data inconsistencies and correctness of assertions. If an error is encountered, the system grinds to a halt, capturing the code for and the location of the call which discovered the error. This technique allows for easy detection of errors and prevents the propagation of errors to the point where their causes become obscure.

However, in practice, the generation of assertions for internal modules may sometimes turn out to be counter-productive, in contrast to the case for protocol-oriented communications software. A protocol is usually implemented at least in part by event-driven FSM's (finite state machines) internal to the protocol monitor. Assertion points and assertions about protocol variables are developed quite naturally (if not easily) for this kind of structure. They tend to occur less naturally in internal software and experience indicates that straining to find them may produce error-prone assertions which may actually slow down the testing/debugging process.

Any or all of the mechanisms described above can be activated or disactivated at compile time using compiler options. For example in PL/M, the text macro feature may be used to declare program lines as test calls or comments, as required.

### 9.4.4.2    Testing/Debugging an X.25 Protocol System (due to Brown)

The environment of the communications software is characterized by asynchronous inputs, and a large variety of errors in the received packets. The inputs to communications software come from two sources. First, packets are being received from the serial communication line, and second, the higher levels of software are placing demands on the communications system to send and receive packets. The error conditions arise from electrical noise on the serial communications line as well as possible software errors and subtle incompatibilities in the protocol at the distant computer.

In order to handle the asynchronous inputs, and also to implement the levels of typical protocols (three in the case of X.25 — ISR, frame, and packet level), the communication software typically contains several processes which also operate asynchronously. The real time combination of input, error conditions, and concurrent execution of processes results in such a variety of possible program flows that it almost defies testing. This problem is solved by performing several phases of tests, each of which places a different emphasis on eliminating accidental errors, verifying functions, and exercising a wide combination of events.

### 9.4.4.2.1    Phases of Testing

A bottom-up approach is followed in testing the communications software. The ISR level is tested first, followed by the frame level, and then the packet level. As each level is tested, it serves as a basis for testing the next level. Thus the test software consists of driver processes for each level and instrumentation software to record and analyze the results.

Most of the testing time for the interrupt service routines is spent learning how to use the hardware. The best source of information is the Hardware Reference Manual.

Both the frame level and the packet level of X.25 are characterized by a single large complicated monitor, and a few small simple processes. The testing effort must thus be devoted to verification of the Frame and Packet Monitors. Because of the conceptual similarity of the frame and packet levels, the testing of each level is done in the same four phases:

1.  Installation of instrumentation software;

2.  Simple loopback test to get the entire software package to execute on the hardware;

3.  Functional test using the Protocol Tester provided by the carrier;

4.  Loopback test with random error generators which create a wide range of situations.

The instrumentation software checks the data structures of a monitor, stopping a test very shortly after an error occurs and prints a history of the activity in the monitor. The instrumentation software is then used as a tool to perform the next three test phases.

In the simple loopback test, the transmitted packets are looped back through a connector to become received packets. The purpose of this test is to eliminate enough accidental errors in the communications driver and instrumentation software that detailed testing of functions can be performed.

The third phase of testing verifies that the communication functions operate as designed and are compatible with the version of X.25 implemented by the carrier. In the case of Bell Canada's X.25, this testing is accomplished interactively using the SNAP Protocol Tester. This is an 8080 based microcomputer which is located at the nearest Datapac node.

In the fourth phase, the transmitted packets are again looped back but this time random errors are added to the received packets. This tests the communication software automatically over a much wider range of situations than can be created interactively with the SNAP Protocol Tester. Here it may be implicitly assumed that the test was successful if no errors occurred in the data structures.

### 9.4.4.2.2  *Instrumentation Software*

The communication software calls subroutines in the instrumentation software to check data structures, record data structures, record program flow information and stop execution of the communications software immediately after an error in the data structures has occurred.

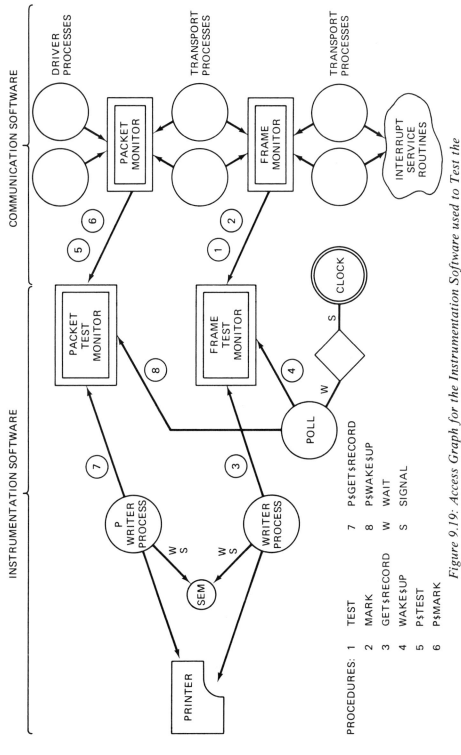

COMMUNICATION SOFTWARE

INSTRUMENTATION SOFTWARE

DRIVER PROCESSES

TRANSPORT PROCESSES

TRANSPORT PROCESSES

INTERRUPT SERVICE ROUTINES

PACKET MONITOR

FRAME MONITOR

PACKET TEST MONITOR

FRAME TEST MONITOR

CLOCK

POLL

P WRITER PROCESS

WRITER PROCESS

SEM

PRINTER

PROCEDURES:
1  TEST
2  MARK
3  GET$RECORD
4  WAKE$UP
5  P$TEST
6  P$MARK
7  P$GET$RECORD
8  P$WAKE$UP
W  WAIT
S  SIGNAL

*Figure 9.19: Access Graph for the Instrumentation Software used to Test the Communications System*

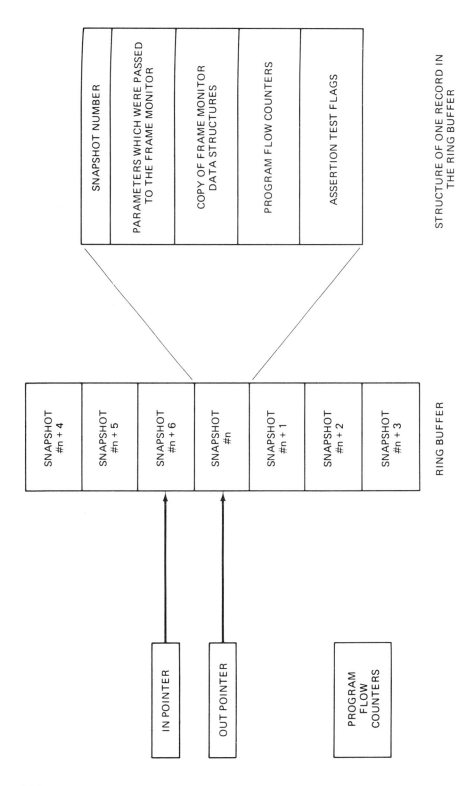

*Figure 9.20: Data Structures of the FRAME$TEST MONITOR*

The structure of instrumentation software is shown in Figure 9.19. Since both the PACKET$TEST and FRAME$TEST Monitor perform identical functions for the packet and frame levels respectively, only the FRAME$TEST Monitor will be described in detail.

The data structures used by the FRAME$TEST MONITOR to record activities in the FRAME MONITOR are shown in Figure 9.20. Each record in the ring buffer is a snapshot of the activities in the FRAME MONITOR. When the FRAME MONITOR calls the TEST subroutine in the FRAME$TEST monitor the following actions occur:

1) The parameters associated with the original call to the FRAME MONITOR are copied into the record.

2) The data structures of the FRAME MONITOR are copied into the record.

3) The program flow counter which was associated with this particular TEST subroutine call is incremented.

4) All the program flow counters are copied into the record.

5) A set of assertions are applied to the data structure of the FRAME monitor, and flags are set in the record to indicate any violations.

6) If there is a violation of an assertion, then all activity in the FRAME monitor is stopped and the WRITER PROCESS is instructed to print out the contents of the ring buffer.

Because the TEST subroutine call performs assertion checks on the data structures of FRAME MONITOR, it can only be called from locations in the FRAME MONITOR where the data structures are known to be consistent. Thus TEST subroutine calls are usually located:

1) at the beginning of the monitor procedure;

2) just before being suspended in the monitor;

3) just before leaving the monitor;

Some of the assertion checks are general in nature and are applied to all TEST subroutine calls, while other assertion checks are made by only certain TEST subroutine calls. The MARK subroutine call to the FRAME$TEST MONITOR simply increments a program flow counter, and is called from locations in the FRAME monitor where the data structures are not expected to be consistent.

The technique of defining assertion checks is difficult in itself and depends on a detailed knowledge of the protocol. As this is not a text on X.25 we omit detailed discussion of the particular checks which may be used for that protocol. Suffice it to say that types of checks are as follows:

— range and order of pointers into packet ring buffers;

— valid packet buffer addresses stored in the ring buffers;

— range of timer and expiry counts;

— FSM states consistent with values of internal variables and error-reporting variables.

To perform the dynamic testing, the following cycle of activities occurs:

1)  The communications software complete with driver processes and instrumentation runs on the hardware until an assertion check is violated.

2)  The ring buffer is automatically printed.

3)  Using the printout of the ring buffer the programmer identifies and corrects the bug.

4)  The code is edited, compiled, linked, etc., and loaded ready to start the test cycle again.

In practice, in the authors' laboratory, a typical test cycle took around 2 hours and usually only the last two records in the ring buffer were required by the programmer to identify the bug. The assertion test flags in the last snapshot indicate which data structures are inconsistent. In practice in many cases the bug occurred in between the second last snapshot (which has no assertion check violations) and the last snapshot (which has at least one assertion check violation). By comparing the copy of FRAME MONITOR or PACKET MONITOR data structures in the last two snapshots, the conditions for the bug as well as the effect of the bug can be seen. The program execution path which contains the bug is marked by the program flow counter which had been incremented between the second last and the last snapshot. With this amount of information in the hands of the programmer there are few dark corners in which bugs may hide.

### 9.4.4.2.3  Conclusions

The experience of the laboratory in the use of program instrumentation in a planned test program for communication software indocates that fast, steady progress in debugging and verification of correctness can be achieved. In the X.25 package it was found that after a test, it usually took about 20 minutes to identify a bug precisely.

In practice, the worst problems are handled by making small improvements in the test bed which either stop the system sooner after the error occurred or yield more information about the error. In the worst case this took about a day, and only occurred a few times.

The exercise of defining assertion checks also forces the designer and programmer to clarify their understanding of the program logic. Sometimes when an assertion check violation occurs during testing, it was the assertion check that is wrong. This generally identifies subtle properties of the protocol or the implementation which had not been noticed in the design or coding stages.

## 9.5  References

The formal equivalence of monitors and message passing is shown in:

HUGH C. LAUER AND ROGER M. NEEDHAM "On the Duality of Operating System Structures" in Proc. Second International Symposium on Operating Systems, IRIA, Oct. 1978 reprinted in Operating Systems Review 13, 2 April 79, pp. 3-19.

The use of message passing as the basis for distributed computing is discussed in:

VARIOUS AUTHORS, Proc. First International Conference on Distributed Computing Systems, Huntsville, Alabma, Oct 1-5, 1979.

JEROME A. FELDMAN "High Level Programming for Distributed Computing" Comm ACM, June 1979, Vol. 22, Number 6, pp. 353-368

E. MANNING, R.W. PEEBLES "A Homogeneous Network for Data Sharing Communciations", Computer Networks 1 (1977), pp. 211-224

C.J. BEDARD, F. MELLOR, W.J. ALDER "A Message-Switched Operating System for a Multiprocessor", Proc. COMPSAC 77, Chicago, Nov. 77

# A PROPOSAL FOR AN
# EFFICIENT POLLING KERNEL
## (due to Paquet)

### A.1   Introduction

This Appendix is concerned with the presentation of a proposal for an efficient polling kernel. Our purpose is not to present a final solution but rather to present in concrete form the development and consequences of some of the ideas expressed in Chapter 9 and in other chapters. Some decisions and assumptions were made prior to designing the kernel:

— All processes on a processor have equal priority.

— Processes are polled cyclically.

— Clock preempts are allowed but not recommended for efficiency reasons.

— Some common memory is available.

— Each CPU owns its own set of kernel routines. Hence, CPU$ID is a parameter which must never be considered.

— The hardware architecture and components are as for the example kernel of Chapter 5.

— The addresses of process description blocks are not passed as parameters to kernel primitives. Instead, a table of process description blocks as well as a pointer to the block describing the currently running process are kept internal to the preempt routine. This is so to optimize preempt efficiency at the cost of some generality.

The structure of the Appendix is as follows. Because kernel related data structures must be passed as parameters to primitives, they are discussed first. Then the primitives are enumerated and described from an external viewpoint. Following that is the description of the internal structure for individual primitives consisting of an algorithm and a set of remarks on the algcrithm. Finally comes a qualitative evaluation of the proposed kernel.

The kernel is presented using PL/M notation.

## A.2   The Data Structures

One of the fundamental characteristics of this kernel is that most of the kernel associated data structures may be declared privately within processes or monitors, and passed as parameters when calling the kernel. This offers the following advantages:

— Initialization is simplified.

— Concurrent operations on these data structures are possible.

— No distinction is made between private and public semaphores (or other structures like mailboxes and condition variables).

— Memory is allocated to these data structures as needed in a dynamic environment rather than at kernel design time.

Moreover, logically related data structures are grouped into single physically related data structures, with the intent to minimize the amount of parameter passing and to maximize the logical simplicity of the various primitives. Protection is required when manipulating individual data structures, and some data structures must reside in common memory.

(a)   *The process Description Block Table (PDBT)*

This table is a structure of process description blocks owned by a preempt routine. Each CPU requires such a table. The table may contain either process description blocks (PDBs) or pointers to PDBs. The proposed kernel uses the first choice for efficiency reasons because it minimizes the levels of indirection. Entries in this table are polled by the preempt routine to find processes which are ready to run.

The PDBT may be located in private memory and needs protection only against clock preemption when manipulated.

(b)   *The Current Process Description Block Pointer (CPDBP)*

This is an address type variable used to store a pointer (to one of the entries in the PDBT) to the process description block of the currently running process. All comments and access rules are as for the PDBT.

(c)   *Process Description Blocks (PDB)*

Process Description Blocks consist of single two byte slots, in which the process's stack pointer (SP) may be stored while the process is inactive. Before a process becomes inactive, its state as well as the information required to ultimately reactivate it is pushed on the STACK.

For systems in which system state information must be saved on each process switch, additional entries in the PDBs may be needed. (Example: memory bank selection at switch time).

Note that the PDBT consists of an array of stack pointers.

Process description blocks may be located in private memory and need protection only against clock preemption when manipulated.

### (d)  *Semaphore Counts (SC)*

For the semaphores defined here, an ADDRESS type variable must be associated with each semaphore. This variable is called the Semaphore's Counts and consists of a WAIT$COUNT (low byte) and a SIGNAL$COUNT (high byte).

All counts for all semaphores are initialized to zero. Each WAIT operation results in the semaphore's wait counter to be incremented modulo 256. Each signal operation results in the semaphore's signal counter to be incremented modulo 256. The criteria by which a process must wait is as follows:

```
INCREMENT WAITCOUNT
IF WAITCOUNT - SIGNALCOUNT = POSITIVE THEN WAIT;
ELSE GO ON;
```

Note that the operation WC-SC = POS is significant only in assembler, and remains significant only as long as waits never exceed signals or signals never exceed waits by more than $256/2\text{-}1 = 127$. This is not believed to be a serious restriction.

Note also that no queues are associated with these semaphores, and hence, that they are not like conventional semaphores. In particular, no dequeueing ever occurs as a result of a SIGNAL operation (it occurs only as a result of a process switch, and then it is not strictly a dequeueing operation, but rather is a selection based on polling).

Semaphore counts need various kinds of protection. For multiple processor systems all write accesses must be critical regions. For waits, the increment wait counts and compare counts must be a critical region. Count comparisons made at preempt time to determine whether a process is ready to go are read only and need not be protected at all. Critical regions may be implemented as one test and set flag per semaphore or as one test and set flag for all semaphores in the system. The later option is adopted for the kernel proposed here.

Semaphore counts are owned by modules other than kernel routines, and passed as parameters. They must be located in common memory only if accessed by processes running on different processors.

### (e)  *Monitor Gates*

Physically gates are semaphores. Logically they are semaphores with special usage. Wait operations need normal semaphore protection. Signal operations are already protected by monitors being critical regions themselves.

### (f) *Compound Gates*

Physically compound gates are two semaphores declared within a single structure as follows:

```
DECLARE COMPOUND$GATE STRUCTURE(
        GATE$WAIT$COUNT           BYTE,
        GATE$SIGNAL$COUNT         BYTE,
        HIGH$PRIOR$WAIT$COUNT     BYTE,
        HIGH$PRIOR$SIGNAL$COUNT BYTE);
```

They are used for more complicated variations of the monitor concept in which both the restarting and the restarted process are allowed monitor code execution following the operation. Neither signal operation needs protection because they already occur within a critical region consisting of the monitor itself.

### (g) *Slots or Condition Variables*

A slot or Condition Variable is the name given to any semaphore used to delay, suspend or stall a process. Physically they are identical to semaphores, but various types of usages are possible as outlined below.

Slots may be multi-count semaphores in which case it is possible to use the same slot to delay several processes in FIFO order.

Slots may also be used as binary slots in which case only a single process may wait on any slot.

Both types of slots may be queued under any desirable discipline in a monitor data structure, thereby allowing medium term scheduling algorithms to be implemented at the application level.

No operations on slot counts need any protection because they already occur within the critical region implemented by the monitor.

### (h) *Link List Mailboxes*

Link List Mailboxes are 6 byte data structures declared as follows:

```
DECLARE LLMB$1 STRUCTURE(
        WAITCOUNT       BYTE
        SIGNALCOUNT     BYTE
        LL$HEAD$PTR     ADDRESS
        LL$TAIL$PTR     ADDRESS)
```

Allowable logical operations on LLMBs are of the PUT$THEN$SIGNAL or WAIT$THEN$GET type. All messages are queued in link listed fashion with HEAD and TAIL pointers maintained in the LLMB data structure.

All operations on LLMBs need protection. Alternatives consist of using a semaphore, a special test and set flag or the same test and set flag as the one used to protect semaphore type data structures. For systems in which link list queueing operations are rare, the first alternative is quite acceptable. LLMBs must be located in public memory only when accessed by processes running on separate processors.

(i) *Binary Mailboxes*

Binary Mailboxes are 6 byte data structures declared as follows:

```
DECLARE BMB$1 STRUCTURE(
        SENDER$WAIT$COUNT          BYTE,
        SENDER$SIGNAL$COUNT        BYTE,
        RCVR$WAIT$COUNT            BYTE,
        RCVR$SIGNAL$COUNT          BYTE,
        MSSG$PTR (OR MESSAGE)      ADDRESS);
```

BMBs are capable of storing one message only at any one time. As such, the producer is blocked on the sender's semaphore until the consumer collects the message. Similarly, the consumer is blocked on the receiver's semaphore if no messages are present at the time of the call.

Protection and location rules are as for Link List Mail- boxes.

## A.3    The Primitives

Kernel primitives are proposed in this section. A complete list of proposed primitives is given in Figure A.1. They are all based on the data structures defined in the previous section, and are described first from an external and then from an internal point of view.

(a) *External Specifications of the Primitives*

```
SIGNAL (SEM$COUNTS$AT);
```

A procedure which will increment the specified semaphore's signal count. Although no dequeueing is performed, this primitive is functionally equivalent (in all respects) to a signal operation on conventional semaphores. The calling process will under no circumstances be preempted as a result of this call.

```
WAIT (SEM$COUNTS$AT);
```

A procedure which will increment the specified semaphore's wait count. This primitive is functionally equivalent to a WAIT operation on conventional semaphores. The calling process may (depending on the current semaphore counts) be preempted as a result of this call.

```
PREEMPT;
```

A procedure which will cause the calling process to be preempted. The preempt routine will subsequently poll the FDBs in its PDBT for potential candidates to run, until one is found. All registers and system state variables will be saved as required. The polling of PDBs will occur according to a built-in (at design time) polling algorithm. A calling process will always be at least temporarily preempted as a result of this call.

```
( 1 )  SIGNAL(SEM$COUNTS$AT);
( 2 )   WAIT(SEM$COUNTS$AT);
( 3 )   PREEMPT;
( 4 )   ENTER(SEM$COUNTS$AT);
( 5 )   EXIT(SEM$COUNT$AT);
( 6 )   DELAY(GATE$COUNTS$AT, SLOT$COUNTS$AT);
( 7 )   CONTINUE (GATE$COUNTS$AT, SLOT$COUNTS$AT);
( 8 )   SUSPEND(GATE$COUNTS$AT, SLOT$COUNTS$AT);
( 9 )   RESTART(SLOT$COUNTS$AT);
(10)    STALL(CONPOUND$GATE$AT, SLOT$COUNTS$AT);
(11)    PROCEED(COMPOUND$GATE$AT, SLOT$COUNTS$AT);
(12)    DEPART(COMPOUND$GATE$AT, SLOT$COUNTS$AT);
(13)    ATTACH$MSSG(LINK$LIST$MAILBOX$AT,   MSSG$PTR);
(14)    DETACH$MSSG(LINK$LIST$MAILBOX$AT,   MSSG$PTR);
(15)    PASS$MSSG(BINARY$MAILBOX$AT, MSSG$PTR);
(16)    ACCEPT$MSSG(BINARY$MAILBOX$AT,   MSSG$PTR).
```

*Figure A.1: Kernel Primitives*

```
ENTER and EXIT (GATE$COUNTS$AT);
DELAY (GATE$COUNTS$AT, SLOT$COUNTS$AT);
CONTINUE (GATE$COUNTS$AT, SLOT$COUNTS$AT);
SUSPEND (GATE$COUNTS$AT, SLOT$COUNTS$AT);
RESTART (SLOT$COUNTS$AT);
STALL(COMPOUND$GATE$AT, SLOT$COUNTS$AT);
PROCEED(COMPOUND$GATE$AT, SLOT$COUNTS$AT);
DEPART(COMPOUND$GATE$AT);
```

These monitor scheduling procedures are proposed as part of the kernel.

```
ATTACH$MSSG(LINK$LIST$MAILBOX$AT, MSSGPTR);
```

This procedure allows messages to be attached to the specified link list mailbox. On entry, MSSGPTR is a pointer to any buffer whose first two bytes remain available to contain a LINK$LIST$PTR. The procedure will queue the buffer, update LL$TAIL$PTR and increment SIGNALCOUNT. The calling process will never be preempted as a result of this call.

```
DETACH$MSSG(LINK$LIST$MAILBOX$AT,    MSSGPTR);
```

This procedure allows messages to be dequeued from the specified link list mailbox. The procedure will wait on WAIT$COUNT, get a message and update LL$HEAD$PTR. On exit, MSSGPTR points to a buffer whose first two bytes are meaningless. The calling process will be preempted only if it must wait for a message to arrive.

```
PASS$MSSG(BINARY$MAILBOX$AT, MSSGPTR);
```

This primitive allows a message to be put in the specified binary mailbox. On entry MSSGPTR is either a pointer to any buffer containing a message, or the message itself if the message is one or two bytes long. Under all circumstances the value of MSSGPTR will be passed to the receiver untouched. The procedure will put the message in the specified mailbox, signal the receiver and wait on the sender's wait count until the message is picked up by the receiver. Hence, the calling process is always preempted as a result of this call.

```
ACCEPT$MSSG(BINARY$MAILBOX$AT,    MSSGPTR);
```

This primitive allows a message to be fetched from the specified binary mailbox. The procedure will wait on the rcvr's wait count, pick up the message and signal the sender's signal count to indicate that the message was picked up. The calling process will be preempted only if it must wait for a message to arrive.

(b) *Internal Structure of the Primitives*

Algorithms for WAIT, SIGNAL, and the internal routine PROCESS$SWITCH were provided in Chapter 9, Section 9.3.2.2, Figure 9.18. The internal algorithms associated with the remaining primitives are discussed from a high level viewpoint in this section.

Notes which further document the algorithms follow the algorithms themselves.

*Preempt*

```
PREEMPT:  PROCEDURE;
          PUSH ALL REGISTERS
          SIMULATE WAIT ON PREEMPT$SEM
          CALL PROCESS$SWITCH
          POP ALL REGISTERS
      END PREEMPT;
```

*Remarks:*

— The preemption semaphore is a special semaphore which at startup is initialized with WAITCOUNT = 0 and SIGNALCOUNT = 1. These counts are never modified. Instead, "SIMULATE WAIT ON PREEMPT$SEM" implies that a process is waiting on that semaphore, without incrementing its WAITCOUNT. This ensures that the next time the calling process's DESCRIPTION$BLOCK is polled the process will be ready to go, because the SIGNALCOUNT will always be greater than the WAITCOUNT.

*Monitor Gates*

```
ENTER:   PROCEDURE (GATE$COUNTS$AT)
             SAME ALGORITHM AS WAIT;
END ENTER;

EXIT:    PROCEDURE (GATE$COUNTS$AT)
             INCREMENT GATE.SIGNALCOUNT;
END EXIT;
```

*Remarks:*

— No protection is needed when signalling a gate. As such exiting a monitor is ultra efficient and does not tie up the test and set flag at all. It becomes an entirely concurrent primitive.

— Should there be enough Test and Set flags or should indivisible test and set instructions be available gates could become test and set flags rather than semaphores for systems with no clock preempts.

— For systems with no timeout or clock preemption, actual suspensions on monitor gates will likely be rare occurrences.

*Delay and Continue*

```
DELAY:   PROCEDURE (GATE$COUNTS$AT, SLOTCOUNTS$AT);
             INCREMENT SLOT.WAITCOUNT;
             INCREMENT GATE.SIGNALCOUNT;
             CALL PROCESS$SWITCH
END DELAY;

CONTINUE:   PROCEDURE(GATE$COUNTS$AT, SLOT$COUNTS$AT);
               IF SLOT.WAITCOUNT > SLOT.SIGNALCOUNT
               THEN INCREMENT SLOT.SIGNALCOUNT
               ELSE INCREMENT GATE.SIGNALCOUNT
END CONTINUE;
```

*Remarks:*

— Note that no loops on test and set flags and no interrupt lockouts exist for these primitives. No protection of any kind is needed (other than reentrancy of course) because all the counts which are manipulated are protected by the monitor gate itself. Hence, these are fully concurrent primitives.

— The SLOT.WAIT count must be incremented before the gate-signal count is incremented in the DELAY primitive to ensure that no other process on any other processor can perform a CONTINUE before the wait count has been incremented. Otherwise a process could "miss" a continue and be delayed forever.

*Suspend and Restart*

```
SUSPEND:  PROCEDURE(GATE$COUNTS$AT, SLOT$COUNTS$AT);
          SAME ALGORITHM AS DELAY
END SUSPEND;

RESTART:  PROCEDURE (SLOT$COUNTS$AT);
          IF SLOT.WAITCOUNT > SLOT.SIGNALCOUNT
          THEN INCREMENT SLOT.SIGNALCOUNT;
END RESTART;
```

*Remarks:*

— The important remarks given for the delay and continue primitives also hold
   true for the suspend and restart primities.

*Stall, Proceed and Depart*

```
STALL:  PROCEDURE (COMPOUND$GATE$AT, SLOT$COUNTS$AT);
        CALL DELAY(GATE$COUNTS$AT, SLOT$COUNTS$AT);
        DECREMENT HIGH$PRIOR$WAITCOUNT;
        CALL WAIT(HIGH$PRIOR$COUNTS$AT);
END STALL;

PROCEED:  PROCEDURE (COMPOUND$GATE$AT, SLOT$COUNTS$AT);
          IF SLOT.WAITCOUNT > SLOT.SIGNALCOUNT
          THEN DO;
                   INCREMENT HIGH$PRIOR.WAITCOUNT;
               INCREMENT SLOT.SIGNALCOUNT
          END;
END PROCEED;
DEPART:  PROCEDURE(COMPOUND$GATE$AT);
         IF HIGH$PRIOR.WAITCOUNT > HIGH$PRIOR.SIGNALCOUNT
         THEN INCREMENT HIGH$PIRORITY.SIGNALCOUNT;
         ELSE INCREMENT GATE.SIGNALCOUNT;
END DEPART;
```

*Remarks*

— These two operators are essential to guarantee corrections for the case where
   DEPART is executed before the unblocked process has a chance to wait on the
   high priority semaphore.

— Allowing the process executing the proceed to terminate monitor code
   execution before any of the proceeded processes re-enter the monitor,
   minimizes the number of required process switches.

*Attach and Detach*

```
ATTACH$MSSG:  PROCEDURE(LINK$LIST$MAILBOX$AT, MSSGPTR);
    CALL QUEUE (LLMB$AT, MSSGPTR);
    INCREMENT LLMB.SIGNALCOUNT;
END ATTACH$MSSG;

DETACH$MSSG:  PROCEDURE (LLMB$AT)   MSSGPTR
    CALL WAIT (LLMB$COUNTS$AT);
    CALL DEQUEUE (LLMB$AT,.MSSGPTR)
    RETURN MSSGPTR
END DETACH$MSSG;
```

*Remarks:*

— QUEUE and DEQUEUE are queueing procedures not shown here.
— Separate protection is needed for these primitives, and is not shown here. Either a special test and set flag or a gate type semaphore could be used. If the latter option is chosen, then these primitives effectively become monitor procedures. There is no apparent reason for which this would be inconvenient.

*Pass and Accept Message*

```
PASS$MSSG:  PROCEDURE (BINARY$MAILBOX$AT, MSSGPTR);
    BMB.MSSG$PTR=MSSGPTR;
    INCREMENT BMB.RCVR$SIGNAL$COUNT;
    CALL WAIT (BMB$SENDER$COUNT$AT);
    END PASS$MSSG;

ACCEPT$MSSG:  PROCEDURE (BMB$AT) MSSGPTR;
    CALL WAIT(BMB$RCVR$COUNT$AT);
    MSSGPTR=BMB.SENDER$SIGNAL$COUNT;
    RETURN MSSG$PTR;
END ACCEPT$MSSG;
```

*Remarks:*

— the same protection applies as for ATTACH and DETACH primitives

## A.4   An Evaluation

Figure A.2 summarizes the major features of a preliminary 8080 assembler implementation of these primitives. A quick glance reveals at least three very encouraging points:

— total memory requirements are minimal;
— the kernel is almost entirely concurrently executable and interrupts are rarely locked out, and then never for longer than for 30 us;

— primitives not involving process switches require less than 50 us of execution time.

The most discouraging point is associated with process switches.:

— Polling preemption is basic to the whole kernel and might not be acceptable for systems in which a large number of processes run on each board.

| PRIMITIVE | WORST EXEC TIME | BEST EXEC TIME | MEM REQs | WORST TAS LOCKOUT | WORST INT LOCKOUT |
|---|---|---|---|---|---|
| WAIT(SEM) ENTER(GATE) | 50 + PREEMPT | 50us | 20 | 25us | 30us |
| SIGNAL(SEM) | — | 41us | 14 | 20us | 25us |
| EXIT(GATE) | — | 16us | 5 | . | . |
| PREEMPT | 50 + n(70)us | 120us | 50 | . | . |
| DELAY SUSPEND | — | 30 + PREEMPT | 12 | . | . |
| CONTINUE | 30us | 25us | 13 | . | . |
| RESTART | — | 25us | 11 | . | . |
| PROCEED | 50us | 22us | 17 | . | . |
| STALL | — | 60 + PREEMPT | 22 | . | . |
| DEPART | — | 35usec | 16 | . | . |

TOTAL  MEMORY    175  BYTES

*Figure A.2: Kernel Statistics*

The kernel itself remains flexible and expandable simply because it is so small and easy to understand. New primitives may be added easily. This is further facilitated by the fact that kernel related data structures are declared external to kernel routines. To illustrate these concepts consider the various message passing primitives discussed in the previous section. Link list and binary mailboxes as well as many other desired data structures may be declared at any time. Primitives to manage them may quickly be coded at any time. They may even be implemented as monitors. In general, the flexibility exists because new data structures and new primitives are independent of old data structures and old primitives, and, hence, extensions are of the addition type rather than of the modification type.

# Index